Topics in Mining, Metallurgy and Materials Engineering

Series editor

Carlos P. Bergmann, Porto Alegre, Brazil

"Topics in Mining, Metallurgy and Materials Engineering" welcomes manuscripts in these three main focus areas: Extractive Metallurgy/Mineral Technology; Manufacturing Processes, and Materials Science and Technology. Manuscripts should present scientific solutions for technological problems. The three focus areas have a vertically lined multidisciplinarity, starting from mineral assets, their extraction and processing, their transformation into materials useful for the society, and their interaction with the environment.

More information about this series at http://www.springer.com/series/11054

Sujay Kumar Dutta · Dharmesh R. Lodhari

Extraction of Nuclear and Non-ferrous Metals

Springer

Sujay Kumar Dutta
Department of Metallurgical and Materials Engineering, Faculty of Technology and Engineering
The Maharaja Sayajirao University of Baroda
Vadodara, Gujarat
India

Dharmesh R. Lodhari
Department of Metallurgical and Materials Engineering, Faculty of Technology and Engineering
The Maharaja Sayajirao University of Baroda
Vadodara, Gujarat
India

ISSN 2364-3293 ISSN 2364-3307 (electronic)
Topics in Mining, Metallurgy and Materials Engineering
ISBN 978-981-10-5171-5 ISBN 978-981-10-5172-2 (eBook)
https://doi.org/10.1007/978-981-10-5172-2

Library of Congress Control Number: 2017945260

Printed on acid-free paper

This Springer imprint is published by Springer Nature
The registered company is Springer Nature Singapore Pte Ltd.
The registered company address is: 152 Beach Road, #21-01/04 Gateway East, Singapore 189721, Singapore

Preface

Extraction of Nuclear and Non-Ferrous Metals is a basic science and has a wide applicability in extraction and refining of metals. This book covers the undergraduate curriculum in **Nuclear Metallurgy and Extraction of Non-Ferrous Metals** as prescribed for degree courses and allied professional courses in Metallurgical and Materials Engineering.

Chapter 1 of this textbook, *Fundamentals of Nuclear Metallurgy*, covers almost all the important basic concepts of **Nuclear Metallurgy** for graduate engineering students. Even postgraduate students, engineers and researchers can brush up their understanding. Chapters 2–6 briefly cover **Nuclear Metals**, their extraction, properties and applications.

Chapters 7–13 discuss **Extractive Metallurgy of Common Metals**. There are detailed discussions about production of common metals (like copper, aluminium, zinc, lead etc.), their properties and applications.

Chapters 14–20 describe the **Extractive Metallurgy of Less Common/Ferro-Alloying Metals.** Their production, properties and applications are discussed in detail. Chapter 21 discusses **Ultra-High-Purity Metals Production.**

Needless to say that although few books are available, none discusses the subject matter of **Nuclear Extractive Metallurgy** in easily understood terms for students. Based on our experience of teaching this subject at undergraduate level for more than three decades, we have made a sincere attempt to cover the topics of **Nuclear Metallurgy** in an easy way.

In spite of taking all possible care, there may be some errors or mistakes left unnoticed. If so, please feel free to interact with us. We poured our long experience into this project, and also collected the information from several sources. We are indebted to one and all, from whose valuable knowledge we have been benefited. Thanks all of them.

We are confident that this textbook will make the subject matter of *Extraction of Nuclear and Non-Ferrous Metals* more simple and easy to understand. We wish to thank the publisher Springer for their continuous support in preparation and publication of this textbook within a short period.

Vadodara, India

Sujay Kumar Dutta
Dharmesh R. Lodhari

Contents

About the Authors

Dr. Sujay Kumar Dutta is a former Professor and Head of the Department of Metallurgical and Materials Engineering, Maharaja Sayajirao University of Baroda, India. He received his Bachelor of Engineering (Metallurgy) from Calcutta University in 1975 and Master of Engineering (Industrial Metallurgy) from Maharaja Sayajirao University of Baroda in 1980. He was awarded Ph.D degree by the Indian Institute of Technology Kanpur, India in 1992. He joined the M. S. University of Baroda as Lecturer in 1981, subsequently being appointed a Reader in 1993 and a Professor in 2001.

Professor Dutta has received several awards, including the Essar Gold Medal (2006) from the Indian Institute of Metals, a Fellowship (2014) from the Indian Institute of Metals, and the IIM Distinguished Educator Award (2015) from the Indian Institute of Metals, Kolkata in recognition of his distinguished service to Metallurgical Education and to the Indian Institute of Metals.

Professor Dutta has published three books along with other authors: *Metallurgical Thermodynamics, Kinetics and Numericals* (2011); *Alternate Methods of Ironmaking (Direct Reduction and Smelting Reduction Processes)* (2012); *Iron Ore–Coal/Coke Composite Pellets* (2013), and *Extractive Metallurgy (Processes and Applications)* is currently in publication. He has also published 120 papers in national and international journals and conference proceedings.

Mr. Dharmesh R. Lodhari is an Assistant Professor at the Department of Metallurgical and Materials Engineering, Maharaja Sayajirao University of Baroda in Gujarat, India. He completed his Bachelor of Engineering in Metallurgy in 1996 and Master of Engineering with specialization in Materials Technology from the Maharaja Sayajirao University of Baroda in 2000. Presently he is pursuing his Ph.D. in the field of new alloy development. He has spent 13 years teaching and researching, has experience in the metallurgical field, and has published numerous technical papers in national and international journals.

Part I
Extractive Metallurgy of Nuclear Metals

Chapter 1
Fundamentals of Nuclear Metallurgy

1.1 Atomic Structure

The operation of a nuclear reactor depends upon various interactions of neutrons with atomic nuclei. In order to understand the nature and characteristics of these reactions, it is desirable to consider briefly some of the fundamentals of atomic and nuclear physics. An atom consists of a positively charged *nucleus* surrounded by a number of negatively charged *electrons*; so the atom as a whole is electrically neutral. Atomic nuclei are combination of two primary particles, namely, *protons* and *neutrons*. Protons and neutrons are jointly named nucleus. The protons carry a single unit positive charge, equal in magnitude to the electronic charge. The neutron is very slightly heavier than the proton and it is an electrically neutral particle carrying no charge. The neutrons are the heart of nuclear energy and they play an important role in the release of atomic energy. All atomic nuclei, with the exception of hydrogen, contain one or more neutrons in addition to protons.

Atomic energy is derived by virtue of the transformation of matter into energy. Calculations show that the conversion of small amount as 1 g of matter yields energy equivalent of 25 million kWh. By contrast, the combustion of the same amount of coal gives the insignificant amount of 0.0085 kWh. The difference in these figures shows the distinction between atomic and chemical energy.

For a given element, the number of protons present in the atomic nucleus is called the *atomic number* and is denoted by **Z**. It is similar to the number of the element in the *Periodic Table* (Fig. 1.1).

e.g. H $\rightarrow$ 1, He $\rightarrow$ 2, Li $\rightarrow$ 3, C $\rightarrow$ 6, O $\rightarrow$ 8, Fe $\rightarrow$ 26, U $\rightarrow$ 92 etc.

The total number of protons and neutrons in an atomic nucleus is called the *mass number* of the element and is denoted by ***A***. Hence, number of neutrons in the atomic nucleus is $(A - Z)$. In general the element (M) can be represented by $_{Z}M^{A}$.

S.K. Dutta and D.R. Lodhari, *Extraction of Nuclear and Non-ferrous Metals*, Topics in Mining, Metallurgy and Materials Engineering,
https://doi.org/10.1007/978-981-10-5172-2_1

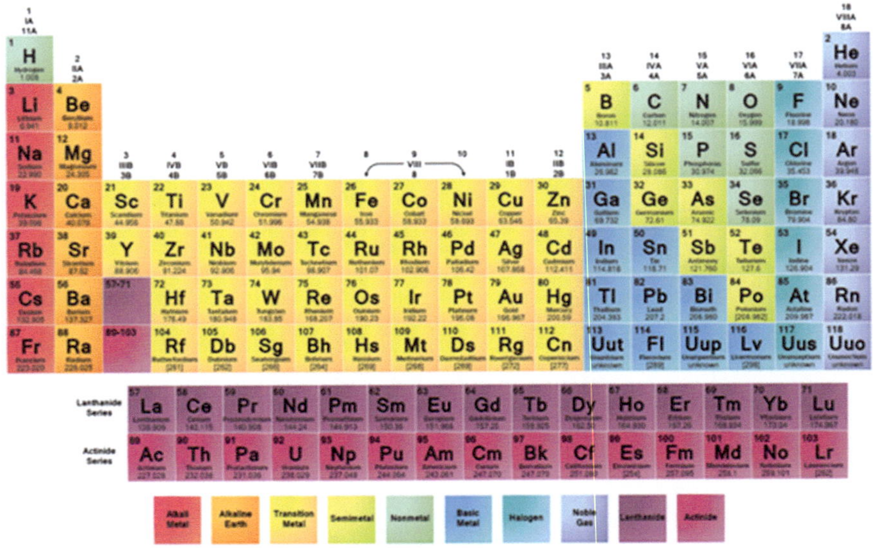

Fig. 1.1 Periodic Table of the elements

1.2 Isotopes

It is the atomic number, i.e. the number of protons in the nucleus, which determines the chemical nature of an element. This is so because the chemical properties depend on the orbital electrons which are surrounding the nucleus; and their number must be equal to the number of protons, since the atom as a whole is electrically neutral.

Atoms with nuclei containing the same numbers of protons (i.e. with the same atomic number, but with different mass numbers) are essentially identical chemically, although they frequently exhibit marked differences in their nuclear characteristics. Such species, having the same atomic number but different mass numbers, are called *isotopes*. Uranium-232 ($_{92}U^{232}$) has three isotopic forms in nature, with mass numbers 234, 235, and 238 ($_{92}U^{234}$, $_{92}U^{235}$ and $_{92}U^{238}$).

1.3 Nuclear Binding Energy

By means of the mass spectrograph, it has been shown that the actual mass is always less than the sum of the masses of the constituent nucleus. The difference is known as *mass defect*, which is related to the energy binding of particles in the nucleus. If m_p, m_n and m_e represent the masses of the proton, neutron and electron respectively, the sum of the masses of the constituents of an atom is $Zm_p + Zm_e + (A - Z)m_n$. Suppose the observed mass of the atom is M, then

$$\text{mass defect} = \left[Z\left(m_p + m_e\right) + (A - Z)m_n\right] - M = Zm_H + (A - Z)m_n - M \quad (1.1)$$

where m_p + m_e = m_H, the mass of the hydrogen atom.

Since m_H = 1.008145 amu* and m_n = 1.008986 amu

(*amu** → *atomic mass unit* is defined as exactly one-sixteenth of the mass of oxygen (O^{16}) atom or one-twelfth of the mass of carbon (C^{12}) atom.)

Therefore,

$$\text{mass defect} = [1.008145\,Z + (A - Z)\,1.008986 - M \quad (1.2)$$

The mass defect is a measure of the energy which would be released, if the individual Z protons and (A − Z) neutrons combined to form a nucleus. Conversely, it is numerically equal to the energy which would have to be supplied to break apart the nucleus into its constituents. Thus, the energy equivalent of the mass defect is called the *binding energy* of the nucleus; i.e. the energy, which holds together the nucleus of an atomic nucleus, is named the *binding energy*.

If m is the decrease in mass accompanying any particular process, then the equivalent amount of energy E released is given by the Einstein equation:

$$E = mc^2 \quad (1.3)$$

where m is mass in grams, c is velocity of light = 2.998 x 10^{10} cm/sec

Therefore,

$$E = m\,x\,8.99x10^{20}\ \text{ergs} \quad (1.4)$$

In the nuclear energy, energies are usually stated in terms of the *electron volt unit*, represented by ev. Since the electronic charge is 1.602 × 10^{-19} C.

$$\text{The electrical energy, } e = vQ\ J, \quad (1.5)$$

where v is in volts, and Q is the amount of electricity in coulomb.

If 1 ev = 1.602 × 10^{-19} J = 1.602 × 10^{-12} erg (since 1 J = 10^7 erg)

1 Mev (million electron volts) = 1.602 × 10^{-6} erg

Therefore,

$$\begin{aligned} E\,(\text{Mev}) &= m\,(\text{gm})\,x\left[\left(8.99\ x\ 10^{20}\right)/\left(1.602\ x\ 10^{-6}\right)\right] = 5.612\,x\,10^{26}\ m \\ &= 5.612\,x\,10^{26}\ x\,1.66\ x\ 10^{-24}m = 931.59\ m\,(\text{amu}) \end{aligned} \quad (1.6)$$

Since 1 amu = 1.66 × 10^{-24} gm

The numerical value of the nuclear binding energy in Mev can be obtained upon multiplying by 931.6, the mass defect in amu, as given by Eq. (1.2). A more useful quantity is the average binding energy per nucleon, which is equal to the total binding energy (BE) divided by the number of nucleons, i.e. by the mass number A; hence from Eqs. (1.2) and (1.6):

$$[BE/A] = [(931.59/A) \text{ x } \{1.008145 \text{ x } Z + 1.008986 \text{ x } (A - Z) - M\}] \tag{1.7}$$

Example 1.1 Determine the binding energy per nucleon in (a) Sn-120 for which M is 119.9401 amu and (b) U-235 for which M is 235.1175 amu.

Solution: (a) The atomic number of Sn is 50, i.e. Z = 50 and A = 120
Since [BE/A] = [(931.59/A) x {1.008145 x Z + 1.008986 x (A − Z) − M}]
Therefore,

$$[BE/A] = [(931.59/120) \text{ x } \{(1.008145 \text{ x } 50) + (1.008986 \text{ x } 70) - 119.9401\}]$$
$$= \mathbf{8.51 \text{ Mev per nucleon}}$$

(b) The atomic number of U is 92, i.e. Z = 92 and A = 235

$$[BE/A] = [(931.59/120) \text{ x } \{(1.008145 \text{ x } 92) + (1.008986 \text{ x } 143) - 235.1175\}]$$
$$= \mathbf{7.59 \text{ Mev per nucleon}}$$

1.4 Radioactivity

There are conventional electrostatic repulsive forces between the positively charged protons. When the atomic number is low, the repulsive force among the protons is small. Hence, the proton–proton, neutron–neutron, and proton–neutron forces are roughly equal; a neutron/proton ratio is close to unity, to be expected for stability. But with increasing atomic number, electrostatic repulsion between the protons, which varies as Z^2, becomes more and more important. Since, the total electrostatic repulsion force between all the protons in the nucleus is proportional to the square of their number, i.e. to Z^2.

In order to maintain stability, the nuclei must now contain an increased proportion of neutrons, so that the attractive neutron–neutron and neutron–proton forces can compensate for the rapidly increasing repulsive forces between the protons. But there is a limit. Consequently, the elements of atomic number 84 or larger have no stable isotopes; although elements 84 (polonium, Po) through 92 (uranium, U) exist in nature, they are unstable and exhibit the phenomenon of *radioactivity*.

Since properties of atomic nuclei depend on the total number of protons and neutrons (nucleons) in the nucleus, isotopes of one and the same element possess different nuclear properties. That is, some isotopes are stable, whereas others are unstable or radioactive.

Radioactive isotopes undergo spontaneous nuclear change (i.e. transformation) at a definite rate in the direction of increasing stability of the element. During this transformation the unstable nucleus emits characteristic particles (i.e. high velocity particles or radiation) and is thereby transformed into a different nucleus which may or may not also be a radioactive element. Isotopes which owe their instability to their high mass numbers emit either positively charged *alpha (α) particles* or negatively charged *beta (β) particles*.

(a) Alpha (α) particles are identical with helium nuclei ($_2He^4$) and consist of two protons and two neutrons. The product (or daughter) nucleus of alpha decay has two protons and two neutrons less than the parent nucleus, so that its mass number is four units less:

$$_ZM^A \rightarrow {}_{Z-2}M^{A-4} + {}_2He^4(\alpha) \tag{1.8}$$

In other words, it produces an element in the Periodic Table (Fig. 1.1) two places to the left of the parent isotope. For example, α disintegration of $_{92}U^{238}$ results in its transformation into $_{90}Th^{234}$:

$$_{92}U^{238} \rightarrow {}_{90}Th^{234} + {}_2He^4(\alpha) \tag{1.9}$$

(b) Beta (β) particles are identical to ordinary electrons. The nucleus itself does not contain electrons, and in radioactive beta decay the electron arises from the spontaneous conversion of a neutron into a proton and an electron:

$$\text{neutron} \rightarrow \text{proton} + \text{electron}\left(\beta\,\text{particle},\ {}_{-1}\beta^0\right) + \text{neutrino} \tag{1.10}$$

The additional neutral particle, with essentially zero mass, called a *neutrino*, carries out some of the energy liberated in the radioactive transformation. In beta decay the daughter nucleus has one neutron less and one proton more than its parent, but the mass number is unchanged; i.e. this produces an element in the Periodic Table (Fig. 1.1) one place to the right of the parent element:

$$_ZM^A \rightarrow {}_{Z+1}M^A + {}_{-1}\beta^0 \tag{1.11}$$

$$\text{e.g. } {}_{92}U^{239} \rightarrow {}_{93}Np^{239} + {}_{-1}\beta^0 \rightarrow {}_{94}Pu^{239} + {}_{-1}\beta^0 \tag{1.12}$$

A neutron is replaced in the nucleus by a proton, so that the neutron/proton ratio decreases, the daughter nucleus will be more stable than its parent, not necessarily completely stable.

(c) In many cases, although not always, radioactive decay is associated with the emission of gamma (γ) rays, in addition to an alpha (α) or beta (β) particles. Gamma (γ) rays are penetrating electromagnetic radiations of high energy, essentially identical with X rays. In fact, the only difference between γ rays and X rays is that the former originate from an atomic nucleus; whereas the later are produced by processes outside the nucleus.

γ rays occur in a radioactive change when the daughter nucleus is formed in what is called an excited state, i.e. a state in which it has a higher internal energy than the normal (or ground) state of that nucleus. The excess energy is then released almost instantaneously as γ radiation. Such an excited nucleus is known as an *isomer of the unexcited nucleus*.

1.5 Rate of Radioactive Decay

The rate of radioactive decay is a value characteristic of a given isotope and independent of the physical and chemical state of the element, at least under the conditions of temperature and pressure. In a given specimen, the rate of radioactive decay at any instant is always directly proportional to the number of radioactive atoms of the isotope under consideration present at that time. Thus, if N is the number of the particular radioactive atoms (or nuclei) present at any time t, the decay rate is given by:

$$(dN/dt) = -\lambda N \tag{1.13}$$

where λ is the radioactive decay (or disintegration) constant of the radioactive species, a measure of its decay probability. Minus sign indicates a decrease in the number of the particular radioactive atoms (or nuclei) i.e. N. This Eq. (1.13) permits a determination of N at a given time t, assuming that there were N_0 parent nuclei at the initial time t_0. By integrate Eq. (1.13):

$$N = N_0 e^{-\lambda t} \tag{1.14}$$

The rate of radioactive disintegration is conveniently expressed by means of a so-called *half-life period* ($t_{1/2}$). It is defined as the time required for the number of radioactive nuclei of a given kind to decay to half its initial value. Assuming $N = 0.5N_0$ and applying to Eq. (1.14):

$$t_{1/2} = (0.6931/\lambda) \tag{1.15}$$

The reciprocal of the decay constant, represented by t_m, is known as *mean life* (or *average life*) of the radioactive species; thus,

$$t_m = 1/\lambda \tag{1.16}$$

The half-life is thus inversely proportional to the decay constant or by Eq. (1.15), directly proportional to the mean life, i.e.

$$t_{1/2} = 0.6931 t_m \tag{1.17}$$

Some of the data for nuclides that are of direct or indirect interest as nuclear fuels are given in Table 1.1. The long half-life is important since it means that there are

Table 1.1 Radioactive characteristics of nuclear fuel isotopes

Nuclide	Radiation	Half-life, years
Th^{232}	α	1.39×10^{10}
U^{233}	α	1.62×10^{5}
U^{235}	α	7.13×10^{8}
U^{238}	α	4.51×10^{9}
Pu^{239}	α	2.44×10^{4}

no appreciable losses in storage over many hundreds of years. It is of interest that, when the fuel species with the shortest half-life, namely Pu^{239}, decays its daughter is U^{235} having a much longer half-life.

$$_{94}Pu^{239} \rightarrow {}_{2}He^{4} + {}_{92}U^{235} \tag{1.18}$$

Since the uranium (U^{235}) is also a nuclear fuel, there is no appreciable loss in usefulness of the material as a source of energy.

Example 1.2 If (dN/dt) = − λN, then prove that: $N = N_0 e^{-\lambda t}$

Solution:

$$(dN/dt) = -\lambda N$$

$$\text{By integrating} \int_{N_0}^{N} (dN/dt) = -\int_{0}^{t} \lambda N$$

$$\text{Or} \quad \int_{N_0}^{N} (dN/N) = -\lambda \int_{0}^{t} dt \quad (\text{since } \lambda \text{ is constant})$$

$$\text{Or} \quad (\ln N - \ln N_0) = -\lambda t \quad \text{or} \quad \ln(N/N_0) = -\lambda t$$

Hence, $\mathbf{N = N_0\, e^{-\lambda t}}$ **Prove**

1.6 Neutron Reaction

In addition to the spontaneous nuclear reactions due to radioactivity, in which one nucleus is changed into another as the result of a particle, there are many nuclear reactions which can be brought about artificially, resulting from the absorption of one particle or another. Among such reactions, those in which a neutron is absorbed are of special interest. Neutrons are normally bound in atomic nuclei, but it is possible to obtain these in the free state. Such free neutrons can interact in various ways with nuclei.

There are three types of neutron-nuclear interactions of importance in the operation of nuclear reactors, namely: (1) scattering; (2) capture; and (3) fission.

(1) *Scattering*: The nucleus absorbs the neutron to form a compound nucleus in an excited state of higher internal energy. This compound nucleus rapidly expels a neutron with a lower kinetic energy than the absorbed neutron, the excess energy remaining on the residual nucleus.

(a) A compound nucleus may emit a particle identical with the one absorbed by the nucleus, the total kinetic energy of the struck nucleus and the bombarding particle remaining unchanged. Such a process is referred to as *elastic scattering*. This kind of reaction neither releases nor absorbs energy, and its only result is a redistribution of the kinetic energy of the collision particles.

$$\text{Reaction}: \ {}_{6}C^{12} + {}_{0}n^{1} \rightarrow {}_{6}C^{12} + {}_{0}n^{1} \text{ or } C^{12}(n,n)C^{12} \tag{1.19}$$

(b) It may happen that the incident particle and the ejected particle are identical, but the total kinetic energy of the colliding particles decreases after the collision. The energy difference is emitted by the compound nucleus as a γ radiation. Such a process is called *inelastic scattering*.

$$\text{Reaction}: \ {}_{92}U^{238} + {}_{0}n^{1} = {}_{92}U^{238} + {}_{0}n^{1}(\text{slow neutron}) + \gamma \tag{1.20}$$

Let E_1 is the total kinetic energy of the neutron and target nucleus before collision, and E_2 is the kinetic energy after collision; if E_γ is the energy emitted as γ radiation, then $E_1 = E_2 + E_\gamma$ (as shown in Fig. 1.2).

(2) *Capture*: The bombarding particle may be trapped in the nucleus. In this case the excitation energy of the compound nucleus is carried away by the γ radiation emitted after absorption of the particle. Such a process is called *capture*. The capture of a neutron by a nucleus followed by the emission of γ radiation.

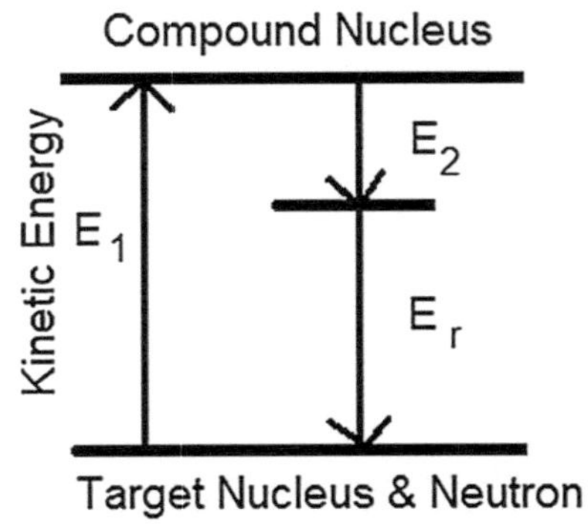

Fig. 1.2 Kinetic energy of the neutron and target nucleus before and after collision

In this reaction the neutron is taken up by the reacting (or target) nucleus in a high energy state. Within a small fraction of a second, the compound nucleus emits the excess energy as radiation, leaving a nucleus differing by one neutron from the target nucleus:

$$_{92}U^{238} + {_0}n^1 = {_{92}}U^{239} + \gamma \tag{1.21}$$

The resulting nucleus, U-239, is radioactive and decays with the emission of a negative beta particle ($_{-1}\beta^0$). Thus:

$$_{92}U^{239} \rightarrow {_{-1}}\beta^0 + {_{93}}Np^{239} \tag{1.22}$$

The product, Np-239 being an isotope of an element of atomic number 93, called neptunium (Np), which does not normally exist on earth. Np-239 is also beta active and decays fairly rapidly.

$$_{93}Np^{239} \rightarrow {_{-1}}\beta^0 + {_{94}}Pu^{239} \tag{1.23}$$

(3) *Fission:* When a compound nucleus attains such a state of excitation that it splits into two parts of lighter masses. This process is called *fission*. When certain nuclei of higher atomic number and mass number capture neutrons, the resulting exited compound nucleus, instead of emitting its excess energy as γ radiation, splits into two lighter nuclei having masses very different from that of the original heavy nucleus.

$$_{92}U^{235} + {_0}n^1 = {_{38}}Sr^{89} + {_{54}}Xe^{145} + \left({_0}n^1 + {_0}n^1\right)\{\text{Fast neutrons}\} + \text{Heat energy} \tag{1.24}$$

When first liberated, free neutrons usually possess high kinetic energies, in the million electron volt range, and so they are called *fast neutrons*. The fission reaction is the basic to the operation of nuclear reactors. Fission occurs only with nuclei of high atomic (and mass) number, and the large value of Z^2, and hence the repulsive force within the nucleus, is an important contributory factor. When fission occurs, the excited compound nucleus formed after absorption of a neutron breaks up into two lighter nuclei, called *fission fragments*.

The important of fission, from the standpoint of the utilization of nuclear energy, lies in two facts: (a) the process is associated with the release of a large amount of energy per unit mass of nuclear fuel; and (b) the fission reaction, which is initiated by neutron, is accompanied by the liberation of neutrons.

1.7 Cross-Sections for Neutron Reactions

The interaction of neutrons with atomic nuclei can be made quantitative by means of the concept of cross-sections. If a given material is exposed to the action of neutrons, the rate at which any particular nuclear reaction occurs depends upon the number of neutrons, their velocity, the number and nature of the nuclei in the specified material. The cross-section of a target nucleus for any given reaction is a property of the nucleus and of the energy of the incident neutron.

Suppose a uniform, parallel beam of I neutrons per cm^2 strikes perpendicularly, for a given time, on a thin layer, dx cm in thickness, of a target material containing N atoms (or nuclei) per cm^3, so that Ndx is the number of target nuclei per cm^2. If C is the number of individual processes, e.g. neutron captures occurring per cm^2. The nuclear cross-section (σ) for a specified reaction is then defined as the average number of individual processes occurring per target nucleus per incident neutron in the beam. Thus:

$$\sigma = [C/\{(\mathrm{Ndx})\,\mathrm{I}\}]\,\mathrm{cm}^2/\mathrm{nucleus}. \tag{1.25}$$

Because nuclear cross-sections are frequently in the range of 10^{-22} to 10^{-26} cm^2/nucleus, it is the general practice to express them in terms of a unit of 10^{-24} cm^2/nucleus, called a *barn*. Thus a nuclear cross-section of 2.7×10^{-25} cm^2/nucleus can be written as 0.27 barn.

1.8 Multiplication Factors

A chain reaction can be maintained only on condition that each fission nucleus produces, on average, at least one secondary neutron; which will cause the fission of another nucleus. This brings us to what is known as the *multiplication factor* (K), the ratio of the number of neutrons of a certain generation (i.e. at a definite stage in the chain reaction) to the number of neutrons absorbed in the preceding generation.

$$\mathrm{K} = (\text{Number of neutrons produced}/\text{Number of neutrons absorbed}) \tag{1.26}$$

A chain reaction is attainable only if K exceeds unity. Assuming that a certain generation consists of 100 neutrons and that K = 1, each succeeding generation will also number 100 neutrons. However, if K = 1.01, then the second generation will have $100 \times 1.01 = 101$ neutrons; the third generation will have $101 \times 1.01 = 102$ neutrons etc. The number of neutrons in each generation increases by $n(K - 1)$ neutrons, where n is the number of neutrons in a given generation.

One of the basic properties of a multiplying system, in which neutrons are being produced by fission, is the infinite multiplication factor (K_∞). It is defined as the ratio of the number of neutrons resulting from fission in each generation to the number absorbed in the preceding generation in a system of infinite size.

In an infinitely large system there is no loss of neutrons by escape (or leakage) and it is only by absorption in fuel, moderator, etc., that neutrons would be removed from the system. The infinite multiplication factor (K_∞) should be unity; the number of neutrons produced in each generation would then be exactly equal to the number of neutrons lost. Thus, a steady state would be maintained with a constant neutron density (number of neutrons/cm^3) and the chain reaction would proceed at a definite constant rate.

In a system of finite size, however, some neutrons are lost by leaking out; the critical condition is then described by means of the effective multiplication factor (K_{eff}), defined as the ratio of the number of neutrons resulting from fission in each generation to the total number lost by both absorption and leakage in the preceding generation. The requirement for critical condition in a finite system is thus $K_{eff} = 1$; in these circumstances a steady state fission chain would be possible. If the conditions are such that $K_{eff} < 1$, then the fission chain would be convergent and would gradually die out. In each generation more neutrons would be lost in one way or another than neutrons are produced by fission, and so the neutron density, and hence the fission rate, would decrease steadily. Such a system is said to be *subcritical.* Finally, if $K_{eff} > 1$, then the fission chain is divergent and the system is known as *supercritical.* More neutrons are produced than are lost in each generation, so that both the neutron population and fission rate increase continuously.

From the definitions of the infinite and effective multiplication factors given above, it is seen that:

$$(K_{eff}/K_\infty) = [\text{Neutrons absorbed}/(\text{Neutrons absorbed} + \text{Neutrons leaking out})] \tag{1.27}$$

1.9 Types of Reactor

Nuclear reactors can be classified according to the kinetic energy of the neutrons which are produced due to the fission reaction in the given reactor.

(1) A reactor system based upon the propagation of slow neutrons with the help of a moderator is called a *slow reactor.*
(2) If a moderator is not used and the chain reaction is caused to be propagated by fast neutrons, the system is called a *fast reactor.*
(3) In some circumstances there may be advantages in using neutrons which have been partially slowed down. A system of this kind is known as an *intermediate reactor*.

Slow reactors are of two sorts, which differ in the disposition of fuel and moderator. In a *homogeneous reactor* the nuclear fuel and the moderator are intimately and uniformly mixed as in a solution of uranyl salt. In a *heterogeneous reactor* the fuel is distributed in discrete units through the moderator.

The slowing down of neutrons or degradation of their energies to low levels is described as a process of moderating them. This is best achieved by surrounding or mixing, the fuel material with a medium consisting of substances of low atomic mass. The medium is known as moderator; graphite, heavy water, or beryllium (Be) are commonly used as moderator.

The principal parts of the slow reactor are as follows:

(a) The *core*, which contains the uranium and the neutron moderator.
(b) The *reflector*, which scatters back neutrons escaping from the core. The reflector is made of the same materials as the moderator;
(c) The *cooling system*, designed to remove the heat released from the reactor core. It consists of pipes through which a coolant is pumped. When passing through the reactor the coolant heats up, transfers its heat to another working medium through a heat exchanger, and then returns to the reactor. Gases (carbon dioxide, air), heavy and ordinary water, and liquid metals (Na, Na-K alloy) can serve as coolants.
(d) The *control system*, designed to control the rate of the chain reaction, and consequently, the power level. This system includes a number of devices: sensing elements, control rods' (b) metal rods containing strong neutron absorbers such as cadmium and boron; and (c) various electronic and electromechanical devices that regulate the position of the control rods.
(e) The protective shield which protects the personnel from radiations emitted in fission and by radioactive fragments.

1.10 Nuclear Fuel and Breeding Reaction

The only fissionable material occurring naturally is uranium-235 (U^{235}), present to the extent of some 0.7% in normal uranium. Out of all the naturally occurring uranium isotopes only U^{235} can be fissioned by slow neutrons, in which it emits several neutrons. It was this isotope that enabled the first uranium fission chain reaction to be carried out. Therefore, U^{235} received the name of *primary nuclear fuel*. Isotopic composition of natural uranium is 0.006% U^{234}, 0.712% U^{235}, and 99.282% U^{238}.

On average, thermal fission of a U^{235} nucleus results in the liberation of 2.5 neutrons:

$$_{92}U^{235} + {_0n^1} \rightarrow {_{38}Sr^{89}} + {_{54}Xe^{145}} + 2.5{_0n^1} + \text{Heat energy}. \tag{1.28}$$

Out of these, one neutron is used to sustain the chain reaction, the remaining 1.5 neutrons being distributed roughly as: 0.9 neutron is captured by U^{238} and 0.6

neutron is partly absorbed by the moderator, the coolant and U^{235} (without fission) and partly escapes from the reactor.

When a U^{238} nucleus captures a neutron, the product is the isotope U^{239}: The latter is a negative beta emitter that decays fairly rapidly (half-life 23.5 min) to form Np^{239}; an isotope of the element neptunium (atomic number 93), which does not exist in nature. Np^{239} is also a negative beta emitter (half-life 2.3 days), and its decay product is the isotope of plutonium, Pu^{239}. The total reaction can be represented by:

$$ {}_{92}U^{238} + {}_{0}n^{1} \rightarrow {}_{92}U^{239} + \gamma \tag{1.29} $$

$$ {}_{92}U^{239} \rightarrow {}_{-1}\beta^{0} + {}_{93}Np^{239} \tag{1.30} $$

$$ {}_{93}Np^{239} \rightarrow {}_{-1}\beta^{0} + {}_{94}Pu^{239} \tag{1.31} $$

Pu^{239} is beta stable, like all heavy isotopes, but it is alpha active. However, its half-life is so long (half-life 24,360 years) that is may be considered practically stable. Pu^{239} is fissionable by slow neutrons and ejects an average of three neutrons. Thus, fission wise, Pu^{239} is similar to U^{235}. Since Pu^{239} does not occur in nature and can be obtained only artificially, it is called *secondary nuclear fuel*.

Secondary nuclear fuel also includes U^{233} which does not occur in nature. It is obtained through a number of thorium transformations resulting from neutron capture:

$$ {}_{90}Th^{232} + {}_{0}n^{1} \rightarrow {}_{90}Th^{233} + \gamma \tag{1.32} $$

$$ {}_{90}Th^{233} \rightarrow {}_{-1}\beta^{0} + {}_{91}Pa^{233} \tag{1.33} $$

$$ {}_{91}Pa^{233} \rightarrow {}_{-1}\beta^{0} + {}_{92}U^{233} \tag{1.34} $$

Although Th^{232} and U^{238} cannot be used directly to maintain a fission chain, they can be converted, by a simple series of nuclear reactions, into the fissionable nuclides U^{233} and Pu^{239} respectively. It is because of the possibility of this conversion that Th^{232} and U^{238} are referred to as *fertile materials*; and their products, after the transformation, are known as *fissile materials* (e.g. U^{233} and Pu^{239}). The mechanisms, by which these transformations occur, are known as *breeding reaction*.

Thus, the burn-up of primary nuclear fuel may be compensated to some extent by the production of secondary nuclear fuel, i.e. Pu^{239} or U^{233}. The ratio of the number of secondary fuel atoms to the number of consumed primary fuel atoms is called the *conversion ratio*. The conversion ratio of U-graphite reactor is 0.9, since the fission of one U^{235} atom yields, on the average, 0.9 Pu^{239} atom.

1.11 Cladding Materials

In heterogeneous reactors nuclear fission is accompanied by the formation of radioactive fission products and dimensional changes in the fuel, such as swelling and cracking. Removal of heat from the reactor core and its subsequent conversion to useful energy is achieved by thermal contact between a flowing coolant and the fuel material. Coolant flow directly past the fuel can corrode and erode the fuel surface, in some instances at a very rapid rate, carrying radioactive products to points external to the reactor core. In addition to preventing attack of the fuel by the corrosive medium, a suitable cladding will prevent radioactive contamination of the primary coolant loop.

The properties of suitable cladding materials can be divided into three categories:

1. Nuclear properties including neutron absorption cross-section.
2. Metallurgical and physical properties including:
   (a) Strength and creep resistance.
   (b) Thermal stability.
   (c) Corrosion resistance.
   (d) Fabricability.
   (e) Thermal conductivity.
   (f) Compatibility with the core.
   (g) Radiation stability.
3. Economic considerations including:
   (a) Availability.
   (b) Cost.

The actual selection of a cladding material for a specified fuel composition and reactor type requires a complex evaluation of the above factors. Zircaloy-2, aluminium (Al) and magnesium (Mg) can be used to clad uranium (U) and thorium (Th) alloys; whereas stainless steel forms a eutectic with these metallic fuels. Stainless steel can be used with water, sodium (Na), organic and gas coolants; whereas Zircaloy-2 is restricted to water and sodium. Aluminium has poor corrosion resistance in high temperature water and organic coolant. Magnesium alloys are used in gas-cooled reactors but not in aqueous, organic, or liquid metal media.

1.12 Radiation Damage

The radiation in a nuclear reactor mostly consist of α and β particles, and γ rays. Apart from that, neutrons, fission fragments, and possibly protons are also radiated. The radiation effect in crystalline solids depends on the structure of material and the nature of the radiation. In a metal, the electrons in the conduction band are able to accept very small amounts of excitation energy, and so a considerable proportion of the energy of the radiation is absorbed for causing electronic excitation. Due to the large number of empty energy states available, the excited electrons rapidly lose their excess energy; this is taken up by the atoms of the metal and appears in the form of heat, i.e. as vibrational energy of the nuclei.

Heavier particles, however, such as protons, neutrons, alpha particles, and fission fragments, effect significant changes in the properties of metals. Due to elastic collisions, these particles may transfer appreciable amounts of energy to the nuclei of a solid. If the amount of energy transferred is enough, it causes the nuclei to be displaced from their normal (i.e. equilibrium) positions in the space lattice. This effect of nuclear radiation is known as *radiation damage*. Properties of materials are changed due to high temperature and intense irradiation. The most important kind of radiation damage in nuclear reactors is that caused by neutrons and γ rays, while nuclear fuel suffers from the additional effects of fission fragments. The radiation damage mechanism consists of the following:

(1) Fast neutrons colliding with atoms of the substance convey a part of their energy to them. In the majority of cases the imparted energy greatly exceeds the atomic binding energy in the molecules. In complex chemical compounds this may result in a rupture of chemical bonds and a change in the chemical composition. In metals it may cause a displacement of atoms from their normal positions in the crystal lattice.
(2) γ rays strip electrons from the electron shells of atoms and impart high velocities to them. Fast electrons strongly ionize the adjacent atoms and break up the chemical bonds of atoms in the molecules.
(3) Fission fragments are very energetic at the beginning of their range and interact with the nuclei of the substance, imparting high velocities to them. When a considerable part of the energy of the fragments is lost in collisions with nuclei, the fragments will not be able to overcome the nuclear repulsive forces any longer, and will spend their energy chiefly in ionizing atoms and rupturing chemical bonds.

These interactions cause substantial changes in the properties of materials undergoing irradiation. Metals lose their mechanical strength, etc.

1.12.1 Atomic Displacement

The fission process produces neutrons and fission fragments having very high kinetic energies (approximately 2 and 100 Mev respectively). These particles collide with surrounding atoms and impart large kinetic energies to them.

Due to an elastic collision between the relatively heavy nuclear particle and an atomic nucleus, the energy transferred to the latter exceeds a certain minimum value (about 25 ev for a metal), the struck (*knocked-on*) atom may be displaced from its normal or equilibrium position in the solid lattice. The displacement of atoms from their positions in the crystal lattice as a result of fast-neutron impact was predicted by E.P. Wigner in 1942. It was consequently been called the *Wigner effect*.

However, if a knocked-on atom is unable to find a vacant equilibrium site, its final location will be a non-equilibrium position. An atom of this kind is supposed to occupy an interstitial site and is known as an *interstitial atom*, or in brief, as an *interstitial*. For each interstitial atom produced by the action of radiation, there must be a corresponding vacant site in the lattice, possibly at some distance away. The net result is a more or less permanent defect in the solid, which may be changed in physical properties.

1.12.2 Temperature Spikes

A struck atom, produced directly by the charged particle or by an energetic knock-on, has sufficient energy to cause it to undergo vibrations of large amplitude without leaving its lattice position entirely. Some of this excess vibrational energy will be rapidly transferred to its immediate neighbours which will, in turn, transfer some energy to their neighbours. The result will be the formation of a limited region in which the atoms have vibration energies in excess of the normal value corresponding to the bulk temperature of the solid. Consequently, the region in which the atoms are in vibrational excited states is referred to as a *temperature spike*.

If the excitation is relatively small, so that few of the vibrational atoms leave their equilibrium sites, the disturbance is called a *thermal spike*. It has been estimated that in such a situation a region containing in the order of a thousand atoms is heated to a temperature of around 1000 °C for a period of about 10^{-10} s.

On the other hand, the vibrational excitation may be sufficient to permit a large number of atoms to leave their lattice sites and more about. As a result of collisions with other atom, a displacement spike will be produced.

1.12.3 Physical Effects of Radiation

The changes produced by radiation on metals and related substances depend on the nature of the materials and the temperature. The effects are smaller at elevated temperatures and in some cases the damage can be removed by raising the temperature. At elevated temperature, atoms can diffuse more rapidly from one site to another through the lattice of a solid; consequently, displaced (interstitial) atoms can more into vacancies and thus restore the normal structure existing prior to irradiation. Under appropriate conditions, many metals exhibit an increase in hardness and tensile strength, but a decrease in ductility as a result of exposure to radiation.

1.13 Reprocessing of Irradiated Fuel

As emphasis in the power-reactor programme shifts from technical feasibility to economic practicality, increasing attention must be given to possible savings in the fuel cycle. Since only a small fraction of fuel is consumed per pass through a reactor, it should be possible to achieve a significant cost reduction by recovery of both fissionable and fertile materials after a fuel element has been subjected to maximum burn-up. The composition and shape of the fuel elements are important in determining the reprocessing scheme.

Where fuel is to be reprocessed for reuse in power reactors, nuclear reactor fuel elements must be discharged for reprocessing long before the fissile and fertile materials are consumed because:

(a) Accumulation of fission products and of the isotopes of heavy elements, which act as neutron poisons, and depletion of the fissile species, e.g. U^{235} and Pu^{239}, can decrease the reactivity to such an extent that the operational requirements of the reactor will no longer be satisfied.
(b) The changes in dimension and shape, which occur with continued exposure and the accumulation of fission products, may set a limit upon the time a fuel element can remain in the reactor.
(c) Accidental circumstances, such as rupture or weakening of the cladding due to thermal shock, weld failures, burn-out, corrosion, or embrittlement, may make it necessary to replace fuel elements.

The objective of the reprocessing operation is to recover in the most economical way as much as possible of these valuable materials in a form in which they can be utilized for reactor (or other) purposes.

In a reactor using natural uranium (or slightly enriched U) as fuel, so that appreciable amounts of U^{238} (99.2%) are present, there will inevitably be some formation of fissile Pu^{239}.

$$U^{238} \xrightarrow{(n,\gamma)} U^{239} \xrightarrow{\beta} Np^{239} \xrightarrow{\beta} Pu^{239} \tag{1.35}$$

If Pu^{239} is allowed to remain in the reactor, it will not only contribute to the fission chain, but it will also capture neutrons to produce the non-fissile Pu^{240}.

$$Pu^{239} \xrightarrow{(n,\gamma)} Pu^{240} \tag{1.36}$$

This will, in turn, undergo the (n, γ) reaction to yield fissile Pu^{241}, and so on.

$$Pu^{240} \xrightarrow{(n,\gamma)} Pu^{241} \xrightarrow{(n,\gamma)} Pu^{242} \rightarrow \ldots. \tag{1.37}$$

The period during which the fuel elements can remain in the reactor will depend upon whether the primary purpose of the reactor is power generation or plutonium (Pu) production. In a power reactor, it may be possible to continue operation for a longer time, since the fissile Pu^{239} and Pu^{241} contribute to the overall neutronic reactivity. On the other hand, if Pu^{239} production is the objective, account must be taken both of the losses and the presence of other Pu isotopes, i.e. changes in the isotopic purity, which will occur with long exposure times.

The reprocessing of irradiated fuel elements generally involves three distinct stages:

cooling; head-end treatment; and separation or extraction.

Two general points must be given attention for reprocessing of spent fuel:

(a) In all stages of fuel element treatment, it is necessary to make absolutely certain that a critical mass of fissile materials is not attained, either in the solid state or in solution.
(b) Because of the intense radioactivity of the spent fuel, even after cooling, special techniques and plant designs are required for carrying out the operations and maintaining the equipment.

1.13.1 Cooling Irradiated Fuel Elements

The purpose of the so-called cooling phase is to permit the decay of various radioactive nuclides present in the spent fuel. These nuclides include fission products and isotopes of elements of high mass number, e.g. U-92 and beyond.

The more significant consequences of cooling are:

(a) The beta and gamma activity is decreased to a level at which radiolytic decomposition of the processing reagents is not significant.
(b) Fission products of short half-life decay almost completely and the reprocessing operation is simplified by the decrease in the number of impurity elements which must be removed from the spent fuel.
(c) Certain heavy isotopes, which cannot be separated chemically from the final products, decay to the point where their radioactivity is no longer a problem.

The manner in which buildup of isotopes of the heavy elements occurs in a fuel consisting of U^{235} and U^{238} is illustrated in Fig. 1.3.

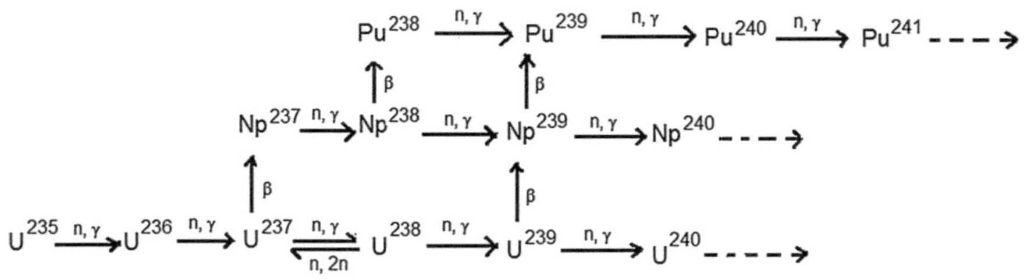

Fig. 1.3 Build-up of isotopes of the heavy elements

1.13.2 Head-End Processes

The purpose of the head-end processes is to prepare the many different types of fuel elements, so that they can all be treated in the same extraction plant. The recovery procedure in common use requires an aqueous nitrate solution of the fuel materials. The head-end processes fall into four categories as follows:

(1) Removal of cladding (or dejacketing) by mechanical means.
(2) Dejacketing by chemical methods.
(3) Dissolution of the fuel but not the cladding.
(4) Complete dissolution of the fuel element.

1.13.3 Separations or Extraction Process

The main purpose of the separation stage of the fuel reprocessing operation is to extract uranium and plutonium in forms that are suitable for further use either as recycle fuels, as feed material in an isotope enrichment plant, or for weapons. In addition, the fission product residues may be processed for the recovery of specific isotopes of scientific or commercial value. There are three methods: (a) aqueous processes like solvent extraction technique; (b) volatility methods depending on the distillation of UF_6; and (c) pyrometallurgical processes.

(a) Solvent extraction technique is a hydrometallurgical process which involves the extraction and recovery of metals using aqueous or organic solutions. Details are discussed at the Sect. 1.14.1.2.
(b) The volatility process of main interest at the present time is based on the fact that UF_6 is volatile at low temperature (sublimation temperature is 54.6 °C), whereas most of the fission products from fluorides do not volatilize readily at all.
(c) In most pyrometallurgical processes the spent fuel is treated in its existing form, so that dissolution or other chemical conversion is eliminated. The basic principle of this process is based on removal of impurities as oxide, and the fuel may remain in the metallic state throughout.

Flow sheet of pyrometallurgical process is shown in Fig. 1.4. The fuel is melted in a refractory crucible of ZrO_2 and held at 1400 °C for several hours. The inert gases (xenon, krypton) and alkali metals (mainly cesium) are volatilized, whereas the Sr, Ba and rare earth metals, which are oxidized more readily than U and Pu, are converted into oxides; these oxides are removed mechanically from the crucible walls. One problem associated with the melting-refining process is the appreciable loss of fissile material carried over together with the impurities to the crucible skull.

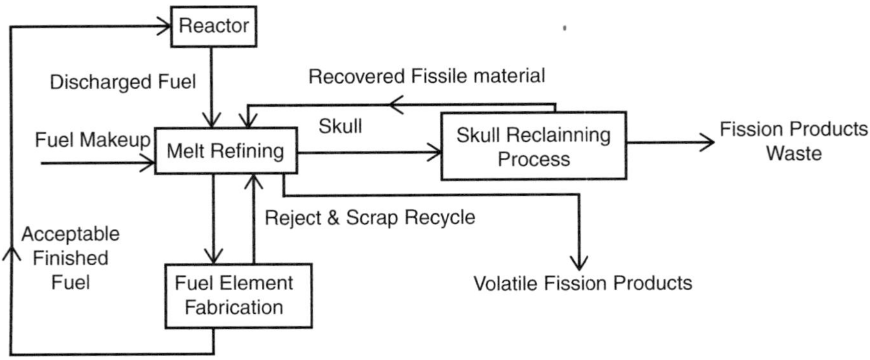

Fig. 1.4 Flow sheet of pyrometallurgical process

1.14 Processing of Nuclear Metals

The development of nuclear reactors for electricity generation has been greatly dependent on the selection and development of suitable materials for the different components with specific functions in the core of the reactor. In addition to the normal properties of good mechanical strength at relatively high temperatures and good corrosion resistance for reliable engineering performance, reactor materials should also satisfy essential criteria with respect to their neutron absorption characteristics and radiation stability.

It is this unusual combination of requirements that has contributed to the emergence of several so-called rare or less common metals. Why are these metals rare or uncommon? Because:

(a) The natural supply in the Earth's crust may be small.
(b) The concentration in accessible deposits may be so low as to require the handling and processing of huge amounts of worthless material in order to extract even small quantities of the desired element even in either compound or elemental form.
(c) The chemical and physical properties of the element may be such that conversion to the elemental form is very difficult.
(d) Even though available, the element may not have enough attractive properties or uses to create a demand for it in competition with other available materials on a cost basis.

Nuclear fuel materials are uranium (U), plutonium (Pu) and thorium (Th). Materials used for fuel cladding are zirconium (Zr) and niobium (Nb); and material used for reflector is beryllium (Be). Minerals of nuclear metals are shown in Table 1.2. Materials for neutron absorption and control are hafnium (Hf) and boron (B). More often, instead of pure metals, their alloys or compounds are used as

Table 1.2 Minerals of nuclear metals in world

Metal	Mineral	Chemical formula	Main location
Uranium (U)	Pitchblende Uraninite Carnotite	U_3O_8 UO_2, UO_3 $K_2O.2UO_3.V_2O_5.xH_2O$	USA, Canada, France, Zaire, Australia, Norway, and Turkey
Zirconium (Zr) & Hafnium (Hf)	Zircon Baddeleyite	$ZrSiO_4$ ZrO_2	USA, Australia, Brazil, and India
Thorium (Th)	Thorianite Brannenite Eschynite	$(Th, U)O_2$ $(U,Ca,Fe,Y,Th)_3Ti_5O_{16}$ $(Ce,Ca,Fe_2,Th)(Ti,Nb)_2O_6$	India, Brazil, Australia, Sri Lanka, Indonesia, Malaysia, South Africa, USA, Korea, Italy, and Egypt
Beryllium (Be)	Beryl	$3BeO.Al_2O_3.6SiO_2$	Zimbabwe, USA, and India
Niobium (Nb)	Columbite Pyrochlore	$(Fe, Mn)(Nb, Ta)_2O_6$ $(Na,Ca)_2(Nb,Ti)(OF)_7$	Zaire, Brazil, Nigeria, Malagasy, Norway, Sweden, Russia

nuclear materials. They are either obtained from the final metal product or from the intermediates in the production flow-sheet of the metal. Table 1.3 summarizes the purity specifications for nuclear metals to satisfy the criteria of neutron economy and to ensure good mechanical and corrosion resistance properties.

By the use of unconventional techniques these rare metals produce in high degrees of purity:

Table 1.3 Specification for nuclear metals (in ppm)

Impurities /	Uranium	Thorium	Zirconium	Hafnium	Beryllium	Niobium
Aluminum	30	–	75	25	100	100
Boron	0.1	0.1	0.5	0.2	3	2
Cadmium	0.1	0.1	0.5	1	2	5
Gadolinium	0.04	0.1	–	–	–	–
Carbon	150	–	250	20	–	100
Hafnium	–	–	100	–	–	–
Iron	40	40	1500	100	50	500
Nitrogen	–	–	50	10	–	300
Oxygen	–	–	1400	90	–	300
Silicon	50	50	100	40	150	300
Tantalum	–	–	–	200	–	1000
Titanium	–	–	50	40	–	500

(1) *Separation process*: Ion exchange and solvent extraction → for separation and purification of the metal from the ore.
(2) *Extraction process*: Metallothermic reduction of pure oxide, or halide intermediates and pyro-vacuum treatment; as well as for purification and consolidation of the final metal product.

1.14.1 Separation Processes

The hydrometallurgical separations are the processes of beneficiation: (a) ion exchange; and (b) solvent extraction.

1.14.1.1 Ion Exchange

The separation by ion-change is achieved because of the differences in the exchange characteristics of different ionic species with the ion exchanger (Fig. 1.5). The technique involves the sequential steps of selective adsorption by the ion exchanger from the feed solution containing a mixture of ions, elution of the adsorbed species from eluate and regeneration of the ion exchange resin for subsequent use.

Ion exchange separations are slow and operations are relatively expensive. However, they are attractive when the component to be separated is present at a very low concentration in the bulk, as in the case of uranium concentration from low grade ores.

$$UO_2(SO_4)_3^{4-} + 4RX = R_4UO_2(SO_4)_3 + 4X^- \tag{1.38}$$

where R is resin and X is chloride/nitrate.

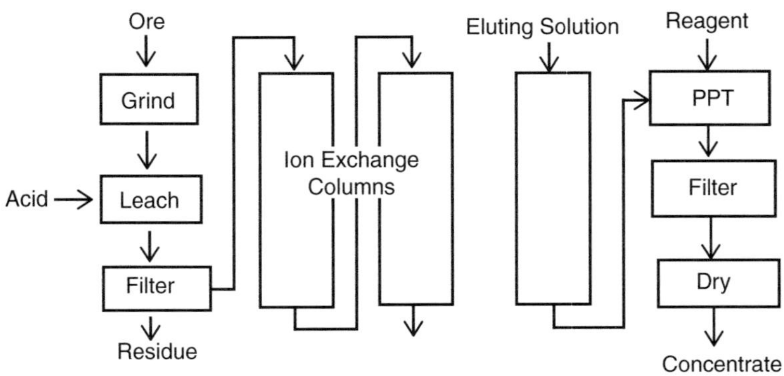

Fig. 1.5 Three columns ion-exchange system

Steps involves in ion exchange area:

(1) Absorbing metal ions from solution.
(2) Removal of the retained ions by replacement reactions using a solution containing proper ion.
(3) Recovering the metal from elute as precipitates.
(4) Filtering and drying to get concentrate.

1.14.1.2 Solvent Extraction

Solvent extraction works on the principle of selective dissolution of one or more components of an aqueous feed solution into a suitable organic solvent. Separation is achieved because of the difference in the distribution coefficients (C_{org}/C_{aq}) of the components in the mixture. Solvent extraction process is readily adapted to high output continuous operation and has advantages like high extraction efficiency and selectivity, ease regeneration of solvent and low installation cost.

The process involves the following steps:

(1) An extraction stage when the solute specie of interest are transferred from the aqueous to the organic phase.
(2) A scrubbing stage when the extraneous impurity solutes transferred to the organic in stage (a) are scrubbed out by contacting the organic with an aqueous phase of proper composition and (b) a final stripping stage → when the metal solute of interest is stripped out from the organic by contacting it with demineralized water or dilute acid.

This process is extensively used in the separation and purification of uranium from ore leach liquors, in the purification of crude thorium nitrate solution and also for separating in quantity, otherwise difficult to separate metal pairs such as Zr-Hf and Nb-Ta. An interesting case is the separation of hafnium from zirconium. These two elements are chemically similar and therefore undergo the same chemical reactions in various processes. They also occur together in leach liquors. However, with tributyl phosphate (TBP)-kerosene as solvent, the solubility of Zr is about an order of magnitude greater than that of Hf. On the other hand, with another organic reagent, NH_4CNS in diethyl ether, more Hf goes to the organic phase. It is, therefore, possible to separate Zr from Hf using solvent extraction.

1.14.2 Extraction Techniques

The intermediates obtained after the hydrometallurgical separation cum purification processes are generally nuclear grade oxides. The next stage in the sequence is the conversion of these intermediates into nuclear pure metals. The currently established processes for nuclear metals production generally employ metallothermic reduction.

1.14.2.1 Oxide and Halide Intermediates

(a) The selection of the extraction route of metal is based on the case of easy removal of oxygen from the metal oxide. This removal of oxygen is feasible by a simple process such as pyro-vacuum treatment. The oxide becomes the preferred intermediate as in the case of Nb and to some extent Hf and Th. The commonly used reductants are Ca and Al for oxides:

$$MO + Ca/Al \rightarrow M + CaO/Al_2O_3 \tag{1.39}$$

(b) In the case of metals where the oxygen removal is not easily achieved from oxide as reduced metals, converting the oxides as intermediates. In such instances, the halides are more convenient intermediates, they yield oxygen free metals by metallothermic reduction. Halide route is used in the case of U, Th, Zr, Hf and Be metals extraction.

Among the halides only fluorides and chlorides are preferred because of their low cost and desirable physical properties. The commonly used reductants for halides are Ca, Mg and Na:

$$MO + 2HF \rightarrow MF_2 + H_2O \tag{1.40}$$

$$MF_2 + Ca \rightarrow M + CaF_2. \tag{1.41}$$

The aggregation of the product metal obtained by the metallothermic reduction of either oxide or halide depends upon the heat of reaction and the densities of slag and metal as follows:

(1) If the consolidated ingot is obtained when the heat of reaction is sufficient to melt both slag and metal, and when the densities of the two are quite different.
(2) Metal in the form of pebbles is obtained when the densities of molten slag and metal are not significantly different.
(3) When the heat of reaction is sufficient only for the metal to get sintered, metal sponge is obtained.
(4) In the case of reductions involving small heat of reactions, metal powders are obtained.

In the last two cases, separation of metal and slag is achieved either by pyro-vacuum treatment or aqueous leaching.

Chapter 2
Uranium

2.1 Introduction

The element uranium (U) was discovered by a German chemist Klaproth in 1789. Metallic uranium was first prepared by Peligot, who showed that the substance discovered by Klaproth was the oxide of uranium. Uranium is the basic (primary) nuclear fuel, since it contains the only naturally occurring fissionable material. In these nuclear applications the element uranium is of interest because of its nuclear properties, providing energy, fission products, and more fissionable material. Uranium may be present in fuel elements as a pure metal, an alloy, or an oxide or other compound. Uranium cannot be handled like the common metals (such as Fe, Cu, Al etc.), because of its chemical reactivity, radioactivity with consequent health hazards, and highly anisotropy. In spite of above difficulties, several tonnes of uranium metal are processed every year in different forms like rod, tube, sheet and plate.

Uranium-235 (U^{235}) is the only naturally occurring fissionable material. Artificial isotopes with mass numbers from 228 to 239 have been prepared. The most important of these is U^{233}, which is prepared from the reaction of neutrons with thorium-232 (Th^{232}).

$$_{90}Th^{232} + {_0}n^1 \rightarrow {_{90}}Th^{233} + \gamma \tag{2.1}$$

$$_{90}Th^{233} \rightarrow {_{91}}Pa^{233} + {_{-1}}\beta^0 \tag{2.2}$$

$$_{91}Pa^{233} \rightarrow {_{92}}U^{233} + {_{-1}}\beta^0. \tag{2.3}$$

Uranium-235 captures a neutron to form uranium-236, which is most likely to undergo fission into a pair of atoms. Depending on the pair fission products, different numbers of neutrons are released.

S.K. Dutta and D.R. Lodhari, *Extraction of Nuclear and Non-ferrous Metals*,
Topics in Mining, Metallurgy and Materials Engineering,
https://doi.org/10.1007/978-981-10-5172-2_2

$$_{92}U^{235} + _0n^1 \rightarrow _{38}Sr^{89} + _{54}Xe^{145}\text{(Fission products)} + 2.5\, _0n^1 \quad (2.4)$$

U^{238} is converted to the fissionable plutonium-239 (Pu^{239}):

$$_{92}U^{238} + _0n^1 \rightarrow _{92}U^{239} + \gamma \quad (2.5)$$

$$_{92}U^{239} \rightarrow _{93}Np^{239} + _{-1}\beta^0 \quad (2.6)$$

$$_{93}Np^{239} \rightarrow _{94}Pu^{239} + _{-1}\beta^0. \quad (2.7)$$

The plutonium and uranium-233 are separated from their respective precursors, uranium and thorium, by chemical means. The separation of the fissionable uranium-235 from natural uranium requires other methods.

2.2 Sources

Uranium is widely distributed in the upper crust of the Earth, albeit somewhat thinly. Uranium contained in the Earth's crust is about 2–4 ppm. It does not occur freely, it is always associated with other minerals. Table 2.1 shows the principal minerals of uranium. Pitchblende and uraninite occur in the form of veins or lodes, the uranium is present in the form of an oxide (U_3O_8); associated with sulphides of silver, lead, nickel, cobalt and iron; and radioactive radium. The later is a degradation product of uranium, the two are always found together. As a secondary mineral, uranium is found associated with vanadates, phosphates, arsenates, etc., the most common is carnotite, a potassium uranyl vanadate, and autunite, a calcium uranyl phosphate, deposits contain averaging up to 1% uranium oxide.

Uranium-bearing ores have been discovered in India at the Singhbhum Thrust Belt (Jaduguda) of Bihar; a few small ones in Rajasthan, in the Himalyan regions of Himachal Pradesh, Uttar Pradesh, and Madhya Pradesh in India. Jaduguda (India) ore contains: 0.08% U_3O_8, 67% SiO_2, 15% FeO. Sea beach sands of Kerala and

Table 2.1 Principal minerals of uranium

Mineral	Composition	Colour	Occurrence
Uraninite	Essentially UO_2 with UO_3 present as a result of oxidation	Black	Norway, South Africa
Pitchblende	U_3O_8	Black	Canada, Congo, India etc.
Carnotite	$K_2O.2UO_3.V_2O_5.nH_2O$ (n = 1–3)	Yellow	USA, South Australia, Congo etc.
Autunite	$CaO.\ 2UO_3.P_2O_5.nH_2O$ (n = 8–12)	Lemon yellow–sulphur yellow	France, USA, etc.

Tamil Nadu contain a useful radioactive mineral, Monazite, which contains thoria, uranium oxides, phosphate and a large percentage of rare earth oxides (9% ThO_2, 0.4% U_3O_8 and 59% RE_2O_3), where RE represent the rare earth metals like scandium (Sc), neodymium (Nd), cerium (Ce), holmium (Ho) etc.

2.3 Extraction of Uranium from Ore

General flow diagram for production of natural and enriched uranium from ore (which contains 0.08% U_3O_8, 67% SiO_2, 15% FeO) is shown in Fig. 2.1. The following steps are required for production of uranium metal from ore:

1. Acid leaching.
2. Ion exchange separation.
3. Production of reactor grade nitrate.
4. Production of uranium dioxide.
5. Metallothermic reduction of uranium compounds.

2.3.1 *Acid Leaching*

The leaching of the uranium valuable from the ore is done by sulphuric acid (H_2SO_4). The uranium goes into solution as uranyl sulphate:

$$2U_3O_8 + O_2 + 6\,H_2SO_4 = 6UO_2(SO_4) + 6H_2O \tag{2.8}$$

The oxygen required in the leaching process is indirectly supplied by the addition of pyrolusite (60–70% MnO_2). The uranyl sulphate present in the lean liquor complexes with excess sulphate ions present in the solution and forms negatively-charged complexes like $UO_2(SO_4)_3^{4-}$, $UO_2(SO_4)_2^{2-}$.

2.3.2 *Ion Exchange Separation*

An ion exchange process using the quaternary ammonium anion exchange resin which exhibits a high selectively for the uranyl sulphate anions is employed to concentrate uranium from the sulphuric acid leach liquors:

$$4RCl + UO_2(SO_4)_3^{4-} = R_4UO_2(SO_4)_3 + 4Cl^-. \tag{2.9}$$

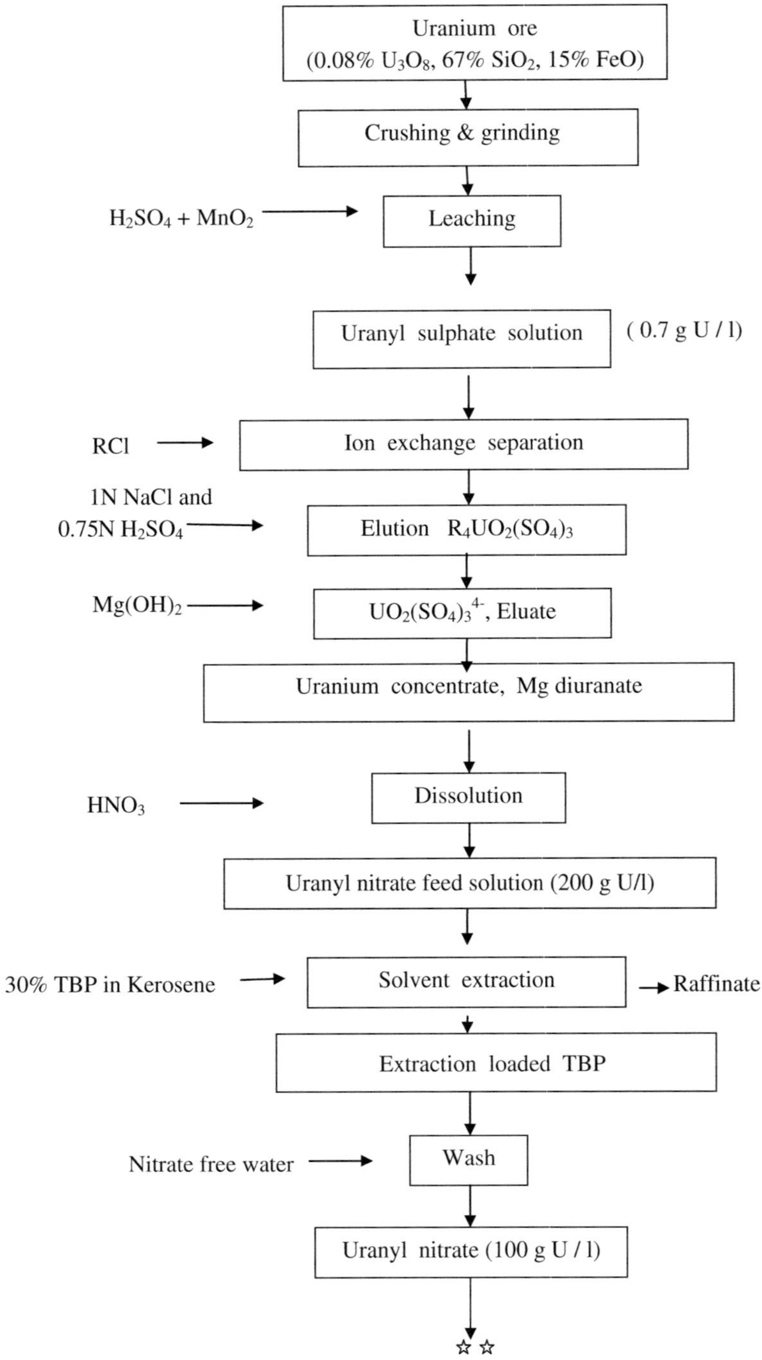

Fig. 2.1 General flow sheet for production of natural and enriched uranium from ore

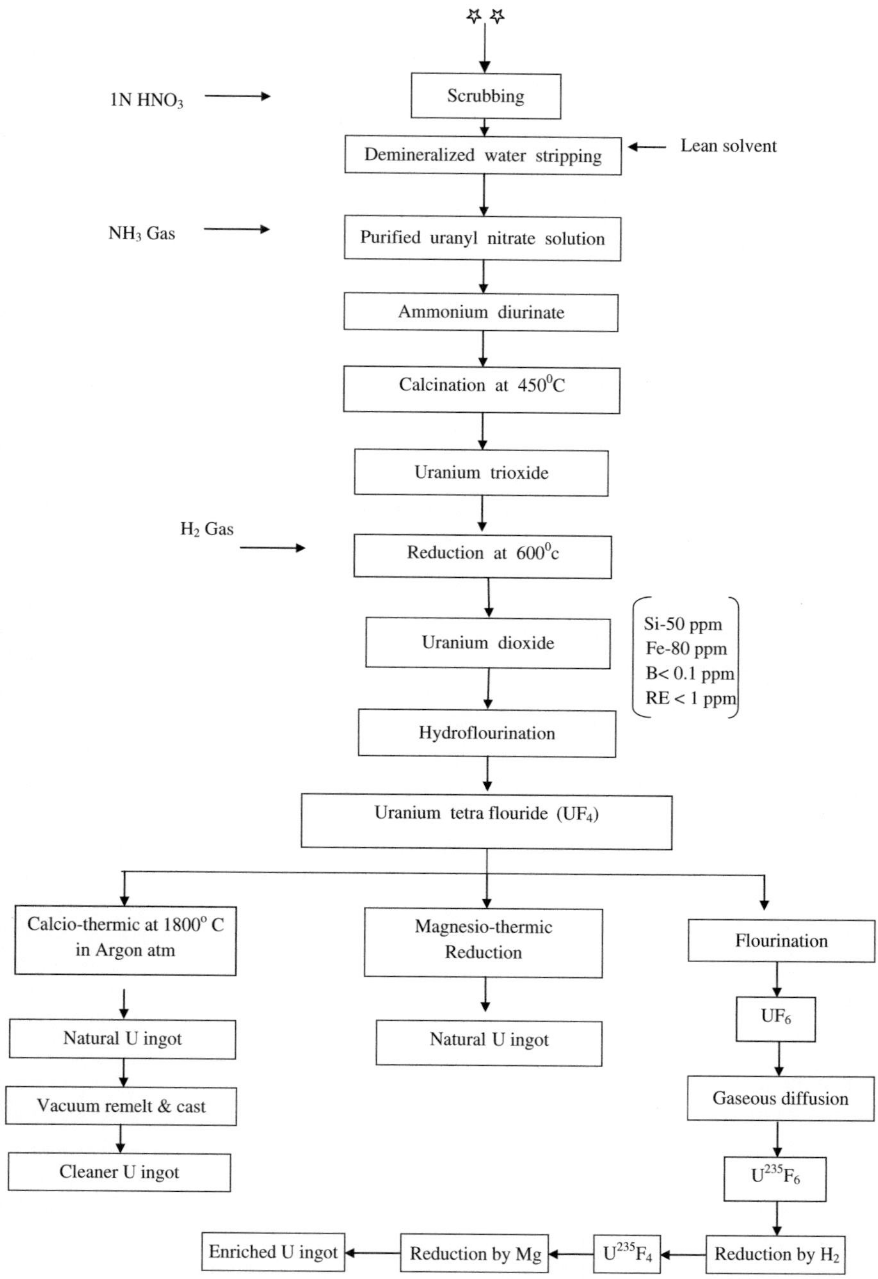

Fig. 2.1 (continued)

The uranium is adsorbed on the anion exchanger in fixed bed ion exchange columns. The adsorbed uranium is collected by passing a strong chloride (NaCl) solution through the saturated resin column:

$$R_4UO_2(SO_4)_3 + 4Cl^- = 4RCl + UO_2(SO_4)_3^{4-} \tag{2.10}$$

Magnesium diuranate is then precipitated from the eluate by the addition of $Mg(OH)_2$.

$$2UO_2(SO_4)_3^{4-} + 6OH^- + Mg^{2+} = MgU_2O_7 + 6SO_4^{2-} + 3H_2O \tag{2.11}$$

2.3.3 Production of Reactor Grade Uranyl Nitrate

The uranium concentrate is first dissolved in nitric acid to obtain a solution of uranyl nitrate $[UO_2(NO_3)_2]$.

$$MgU_2O_7 + 4HNO_3 = UO_2(NO_3)_2 + Mg(OH)_2 + H_2O \tag{2.12}$$

This compound is soluble in various organic solvents such as diethyl ether, methyl isobutyl ketone (hexone), and tributyl phosphate (TBP). TBP selectively forms neutral addition complex $UO_2(NO_3)_2.2TBP$ with uranyl nitrate.

$$UO_2^{2+}.6H_2O_{(aq)} + 2NO_{3(aq)}^- + 2TBP_{(or)} = UO_2(NO_3)_2.2TBP_{(or)} + 6H_2O \tag{2.13}$$

When the loaded TBP is washed with nitrate free water pure uranyl nitrate is obtained in the aqueous phase.

2.3.4 Production of Uranium Dioxide

This aqueous solution is again dissolved in nitric acid for scrubbing (i.e. cleaning) and stripping (i.e. removing) by water to form purified uranyl nitrate solution. Then by passing ammonia gas, ammonium diuranate $[(NH_4)_2U_2O_7]$ is precipitated.

$$2UO_2(NO_3)_2 + 2NH_3 + 3H_2O = (NH_4)_2U_2O_7 + 4HNO_3 \tag{2.14}$$

Now $(NH_4)_2U_2O_7$ is calcinated at 450 °C to obtain uranium trioxide (UO_3).

$$(NH_4)_2U_2O_7 = 2UO_3 + 2NH_3 + H_2O \tag{2.15}$$

The uranium trioxide is then reduced by hydrogen at 550–600 °C to form uranium dioxide. This uranium dioxide contains 50 ppm Si, 80 ppm Fe, < 0.1 ppm B, and < 1 ppm RE.

$$UO_3 + H_2 = UO_2 + H_2O \tag{2.16}$$

2.3.5 *Reduction of Uranium Compounds*

Pure uranium metal is difficult to prepare because the element has affinity for other elements such as oxygen, halogens, nitrogen and carbon. So, the reduction has to be performed in closed systems to avoid atmospheric contamination. The large negative free energy of formation of uranium dioxide (UO_2) [−514.6 kJ/g-atom of O_2 at 25 °C] shows the need for strong reducing agents if UO_2 is to be used as starting material in the preparation of the metal. Hydrogen would require a very high H_2/H_2O ratio in the gas mixture and cannot be considered a practical reducing agent. Reduction with carbon requires a vacuum and leads to contamination by carbide formation.

Thermodynamically, uranium dioxide (UO_2) can be reduced to uranium metal by the reducing agents calcium and magnesium. But the refractory oxides product does not melt in the process and uranium is obtained dispersed through the slag matrix as fine powder. The product is inconvenient to handle, impure and this method is not used to produce uranium.

Halides are more suitable than oxides as starting materials for metal preparation. Additional heat is evolved, and the halide by-product has a lower melting point. This halide therefore melts and permits the dense uranium metal to settle at the bottom. Fluoride reduction is therefore universally employed to obtain uranium metal. The fluoride can involve high heat of reaction yielding oxygen free consolidated uranium metal.

2.3.5.1 Preparation of Uranium Tetrafluoride

The uranium tetrafluoride is less hydroscopic and less volatile than the chloride and it is the preferred halide intermediate. The uranium tetrafluoride is prepared by hydroflorination of uranium dioxide:

$$UO_2 + 4HF = UF_4 + 2H_2O \quad \Delta H_{298} = -266.1\text{kJ}. \tag{2.17}$$

The reaction is highly exothermic and continuous hydroflorination is presently carried out in magnesium-lined inclined rotary reactors. The reaction is carried out at 600 °C and the UO_2 and HF feed rates are adjusted to obtain complete conversion of UO_2 to UF_4.

2.3.5.2 Reduction of UF_4

(i) Calciothermic reduction of UF_4: The reduction with calcium is simpler because of the high heat of reaction which raises the charge temperature to as high as 1800 °C.

$$UF_4 + 2Ca = U + 2CaF_2 \ \Delta H_{298} = -575.7\,kJ. \tag{2.18}$$

Melting points of U and CaF_2 are 1132 °C and 1418 °C respectively, both are molten stages at 1800 °C; and so the slag-metal separation will be good. Calcium has a high boiling point (1494 °C) and hence the reaction takes place at normal pressure.

High purity UF_4 powder is intimately mixed with high purity vacuum distilled calcium and the mixture is charged in a tight stainless steel bomb reactor which is initially lined with dry CaF_2. Calciothermic reduction of UF_4 is done in an argon atmosphere. Uranium ingots obtained are usually remelted in a vacuum electric furnace and cast into cleaner ingots.

(ii) Magnesiothermic reduction of UF_4: Magnesium is easier to get in high purity. Although the heat of the reaction is low and external heat supply is required for good slag-metal separation, Mg is still preferred for uranium production.

$$UF_4 + 2Mg = U + 2MgF_2 \ \Delta H_{298} = -372.4\,kJ. \tag{2.19}$$

The boiling point of magnesium (1105 °C) is lower than the melting point of uranium (1132 °C). Hence, a high pressure develops during the reaction and a pressure vessel is required to carry out the reduction. To supplement the heat requirement of the process, potassium chlorate is added as a heat booster. The reduction vessel is lined with MgF_2.

(iii) Production of UF_6: UF_4 is reacted with fluorine gas to form UF_6:

$$UF_4 + F_2 = UF_6 \tag{2.20}$$

By the gaseous diffusion to form $U^{235}F_6$, then reduced by hydrogen:

$$U^{235}F_6 + H_2 = U^{235}F_4 + 2HF \tag{2.21}$$

Then further reduction is done by magnesium to get enriched uranium:

$$U^{235}F_4 + 2Mg = U^{235} + 2MgF_2. \quad (2.22)$$

2.3.6 *High Purity Uranium Metal*

Small ingots of massive pure uranium are made by vacuum melting of high purity crystals of uranium in uranium oxide crucibles. These crystals are prepared by electrolysis in an argon atmosphere. The anode is ordinary vacuum cast uranium and the cathode is molybdenum (Mo). The electrolyte is a fused salt, comprising a eutectic mixture of LiCl, KCl and 30% UF_4 at 400–450 °C.

2.4 Properties

The uranium metal possesses a silver-grey luster when freshly prepared, but tarnishes readily on exposure to air owing to oxide formation. Uranium is a dense, moderately hard metal and has attractive properties such as high specific gravity and thermal conductivity. The atomic mass of uranium is 238.07. The radiation energy and half-life of natural uranium are shown in Table 2.2. The specific gravity of uranium depends on the amount of carbon and other low atomic number impurities present in uranium. The specific gravity for high purity uranium metal is 19.05 and for wrought uranium metal (variable purity) is 18.7–19.1. The melting point and boiling point of uranium are 1132 °C and 3813 °C respectively. It possesses a tensile strength of 63 kg/mm^2, with reasonable ductility (10% elongation). The strength rapidly deteriorates above 400 °C. Thermal conductivity (0.0645 cal/cm °C) and electrical resistivity (30 μΩ cm at 25 °C) of uranium are increased with increasing temperature. Electrical resistivity show discontinuities at the phase transformation temperature, but thermal conductivity is continuous. Uranium can be cast and fabricated into desired shapes by conventional means such as rolling, extrusion and drawing. The metal is, however, very reactive and oxidizes at moderately high temperatures and as a consequence must be protected from air during fabrication.

Table 2.2 Radiation energy and half-life of natural uranium

Mass no.	Radiation			% in natural uranium
	Type	Energy (Mev)	Half-life (years)	
234	α	4.76	2.48×10^5	0.0056
235	α	4.39	7.13×10^8	0.718
238	α	4.18	4.51×10^9	99.276

Table 2.3 Phase transformation in uranium

Transformation	Temperature (°C)	Volume change (ΔV/V) %	ΔH, kJ/g-atom
$\alpha \rightarrow \beta$	667	1.06	2.93
$\beta \rightarrow \gamma$	775	0.736	4.81
$\gamma \rightarrow$ Liquid	1132	5.0	19.66

Uranium has three allotropic forms (between room temperature and its melting point): (a) orthorhombic—α phase which is stable from room temperature to 667 °C; (b) tetragonal—β phase, stable from 667 to 775 °C; and (c) BCC—γ phase, stable from 775 to 1132 °C. Phase transformation in uranium takes place due to temperature change as shown in Table 2.3.

Hardness: The vickers hardness number (VHN) is commonly used as the scale for hardness of uranium. At room temperature, a hardness of about 220 VHN is found for uranium which has been either cast or α—rolled and annealed. The hardness falls sharply as the temperature increases. The rate of decrease becomes more rapid at 350–400 °C, and a sharp rise in hardness due to transformation from $\alpha \rightarrow \beta$. The hardness of β decreases gradually with temperature, transformation to γ gives a very soft phase like lead (Pb) at room temperature.

Alloys: Uranium alloys are formed due to: (a) to obtain a finer grain size; (b) to improve mechanical properties; (c) to change its properties so that it can be fabricated with other metals; (d) to improve corrosion resistance; (e) to improve resistance to radiation damage; (f) to lower the melting point for liquid metal fuels or to improve castability for cast fuels.

Uranium can be alloyed with metals like Zr, Si, Al, Nb, Mo, Cr, V and Ti.

(i) Metals, those permitting β—U at room temperature, are: V, Nb, Cr, Mo, Mn.
(ii) Metals, those permitting γ—U at room temperature, are: Nb, Mo, Zr.

Certain alloying elements, when added to uranium, can increase its corrosion resistance in high-temperature water, presumable through the formation of denser and more tenacious oxide films than those formed on unalloyed uranium. The alloys may be grouped into four classes:

1. Supersaturated α phase (martensitic) alloys, e.g. U—1.5 wt% Nb—5 wt% Zr.
2. Metastable γ phase alloys, e.g. U—7 to 15 wt% Mo or 9 to 20 wt% Nb.
3. Intermetallic compounds or intermediate phases, e.g. ε phase—stoichiometric composition of U_3Si.
4. Uranium containing relatively small amounts of alloying elements such as Zr, e.g. U—2 wt% Zr, U—5 wt% Zr, U—7.5 wt% Zr etc.

Corrosion: Uranium reacts readily with air, oxygen, nitrogen, hydrogen, water and many other media. Ordinarily the rates of reaction increase rapidly with increasing temperature. Uranium reacts readily in air at room temperature, forming a layer of uranium oxide which at first is adherent but which eventually becomes loose and powdery. At elevated temperature the rate of reaction with air becomes faster.

$$U(s) + O_2(g) = UO_2(s) \tag{2.23}$$

$$U(s) + 3/2H_2(g) = UH_3(s) \tag{2.24}$$

$$U(s) + C(s) = UC(s) \tag{2.25}$$

Uranium reacts with water to form uranium oxide and release of hydrogen gas:

$$U(s) + 2H_2O(l) = UO_2(s) + 2H_2(g) \tag{2.26}$$

2.5 Applications

For use in atomic energy reactors (atomic energy is derived by virtue of the transformation of matter into energy) the uranium metal is cast into rods approximately 2.5 cm in diameter using vacuum melting technique in a high frequency electric furnace.

Since the discovery and isolation of plutonium and uranium-235 took place during a war period, it was natural that attention should have been directed to the military angle. As early as 1940, before the separation of U-235 had been achieved or the existence of plutonium had ever been imagined, scientists were giving attention to the possibility of producing an atomic weapon of unprecedented power.

The rapid growth and development of the nuclear industries has been in many ways unique. Today, many countries are either planning or already developing nuclear power programmes. Britain began operating the world's first industrial-scale nuclear power station in 1956.

Chapter 3
Plutonium

3.1 Introduction

Plutonium (Pu) is the first synthetically produced element. Metallurgy of plutonium differs from that of other elements. Plutonium is unique amongst metals both in its origin and behaviour. It is a synthetic material in the sense that it does not occur naturally as such but is artificially produced in nuclear reactors from uranium-238 by neutron absorption. In 1942, the first pure compound of plutonium was prepared. In 1943, scientists at the University of Chicago made the first reductions of plutonium. Minute beads (weighing less than 50 μ gm) of metallic plutonium were produced. Since then, several laboratories in the USA, Canada, Great Britain, the USSR and France have contributed greatly to the metallurgical knowledge of plutonium. Physical behaviour of metallic plutonium is characterized by an unusual allotropic behaviour; there are six crystallographic forms which exist below melting point (640 °C): α, β, γ, δ, δ' and ε. It is one of the most toxic substances known to us, the permissible contamination levels for plutonium being the lowest for any of the radioactive elements. But peaceful application of plutonium as a source of energy in power reactors of the future is estimated to be capable of multiplying the energy contained in the world's uranium reserves by a factor of more than 100.

The handling of plutonium requires special facilities and extreme caution owing to: (a) its fissionable nature, which necessitates critical mass considerations at all stages in handling; (b) health hazards due to its natural radioactivity and chemical affinity for the body.

There are 15 isotopes (at varies from 232 to 246) of plutonium. In practice, the most important isotope of plutonium is Pu-239. Because, it undergoes fission with both high and low energy neutrons, and can be produced in appreciable amounts, it is of prime importance in the release of nuclear energy.

Although plutonium as a fissionable material has been used principally in nuclear bomb production, this element will undoubtedly become of increasing importance in nuclear power generation. As the consumption of world supplies of fissionable U^{235}

S.K. Dutta and D.R. Lodhari, *Extraction of Nuclear and Non-ferrous Metals*,
Topics in Mining, Metallurgy and Materials Engineering,
https://doi.org/10.1007/978-981-10-5172-2_3

for power generation increases, the fast breeder reactor will become of the utmost importance, in as much as in such a reactor the production of Pu^{239} from U^{238} will be greater than the amount of fissionable fuel consumed. Under fast neutron irradiation the fission of a Pu^{239} nucleus yields 2.9 neutrons, on average, as compared with 2.5 neutrons per fission of U^{235}. This means that if a fast reactor is fueled with Pu^{239}, at least 1.5 Pu atoms can be produced from natural uranium for each plutonium atom consumed, thus aiding the conservation of nuclear fuel reserves.

3.2 Sources

Plutonium is formed as the result of radioactive capture of neutrons by U^{238} and subsequent two stages β decay of the intermediate products. In nuclear reactors fuelled with natural uranium (containing 0.71% U^{235}), neutrons produced by the fission of U^{235} are captured in U^{238} (99.2%) to yield Pu-239 by the so-called *pile-reaction*.

$$_{92}U^{235} + {_0}n^1 \rightarrow {_{38}}Sr^{89} + {_{54}}Xe^{145}(\text{Fission products}) + 2.5\ {_0}n^1 \tag{3.1}$$

$$_{92}U^{238} + {_0}n^1 \rightarrow {_{92}}U^{239} + \gamma \tag{3.2}$$

$$_{92}U^{239} \rightarrow {_{-1}}\beta^0 + {_{93}}Np^{239} \tag{3.3}$$

$$_{93}Np^{239} \rightarrow {_{-1}}\beta^0 + {_{94}}Pu^{239} \tag{3.4}$$

Isotopes of Pu^{232} to Pu^{246} have been synthesized, but at present Pu^{239}, with its half-life of 24,360 years, is the most important in the production of nuclear fuels and bombs. Although plutonium does occur naturally in minute quantities, for all practical considerations it is a synthetic element. In 1942, scientists established the presence of Pu^{239} in pitchblende (U_3O_8) and other uranium minerals in concentrations of about 1 part in 10^{14} (Table 3.1). No isotopes of plutonium other than Pu^{239} have been found in nature.

Table 3.1 Plutonium content in U and Th ores

Sr. no.	Ore	U contain %	Ratio [Pu^{239}/ore (wt)] x 10^{12}	Ratio [Pu^{239}/U] x 10^{12}
1.	Canadian Pitchblende (U_3O_8)	13.5	9.1×10^{-1}	7.1
2.	Belgian Congo Pitchblende	38.0	4.8	12.0
3.	Colorado Pitchblende	50.0	3.8	7.7
4.	Belgian Congo Pitchblende Concentrate	45.3	7.0	15.0
5.	Brazilian Monazite (ThO_2, U_3O_8, RE_2O_3)	0.24	2.1×10^{-2}	8.3
6.	North Carolina Monazite	1.64	5.9×10^{-2}	3.6
7.	Carnotite ($K_2O.2UO_3.V_2O_5.nH_2O$) n = 1–3	10.0	4×10^{-2}	0.4

3.3 Extraction of Plutonium

A flow diagram of plutonium extraction is shown in Fig. 3.1. Production of Pu-239 begins with the reactor operation of any nuclear fuel element that contains U^{238} isotope. Since there are numerous types of uranium fuelled reactors, both the nuclear conditions and the metallurgical environment under which plutonium formation occurs will vary considerably. These considerations affect both the nuclear qualities of the resulting plutonium and the details of the processes for separation of the product from the parent fuel.

Plutonium can exist in a number of isotopic forms. The product of a nuclear reactor is actually a mixture of isotopes in which the percentage of each isotope present is a function of period of irradiation, temperature of irradiation, neutron-flux energy distribution, and isotopic content of the starting material and radioactive half-lives of the isotopes produced. Under short periods of irradiation, a relatively pure Pu^{239} is formed. The thermal neutron capture cross-section of Pu^{239} is about 300 barns, which is slightly less than half the fission cross-section. In addition, the capture to fission ratio at the Pu^{239} neutron resource energies is also high. Consequently, when a fuel element containing U^{238} or Pu^{239} is subjected to prolonged exposure in a thermal reactor, appreciable quantities of Pu^{240} are formed. The thermal neutron capture to fission ratio of Pu^{240} is very high, and, therefore, this isotope is in turn converted to Pu^{241} and higher isotopes by successive neutron capture.

With the low uranium burn-up normally achieved in thermal reactors, the isotopic content of the resulting plutonium may run approximately 94.6 at.% Pu^{239}, 5 at.% Pu^{240}, 0.3 at.% Pu^{241}, and with negligible quantities of Pu^{242} present.

3.3.1 *Separation of Plutonium*

The major problem in recovering plutonium is its separation from the associated uranium and fission products. The amounts of plutonium present in irradiated uranium are usually of the order of only a few grams of plutonium to several tonnes of uranium. Wet chemical separation processes were developed to accomplish stripping of Al-cladding, dissolution of the uranium, and separation of the plutonium from the uranium and accompanying fission products. Plutonium is obtained from these processes in the form of a salt or an oxide, which is then reduced to metallic plutonium by conventional means.

A number of satisfactory procedures have been developed for the separation and recovery of plutonium. Classification of methods includes: (a) precipitation; (b) organic solvent extraction; (c) ion exchange; and (d) pyrometallurgy.

Choice of a method or combination of methods depends largely on the economics of the particular situation. The first step in recovery of plutonium from irradiated natural uranium fuel elements is to store the elements under water for a

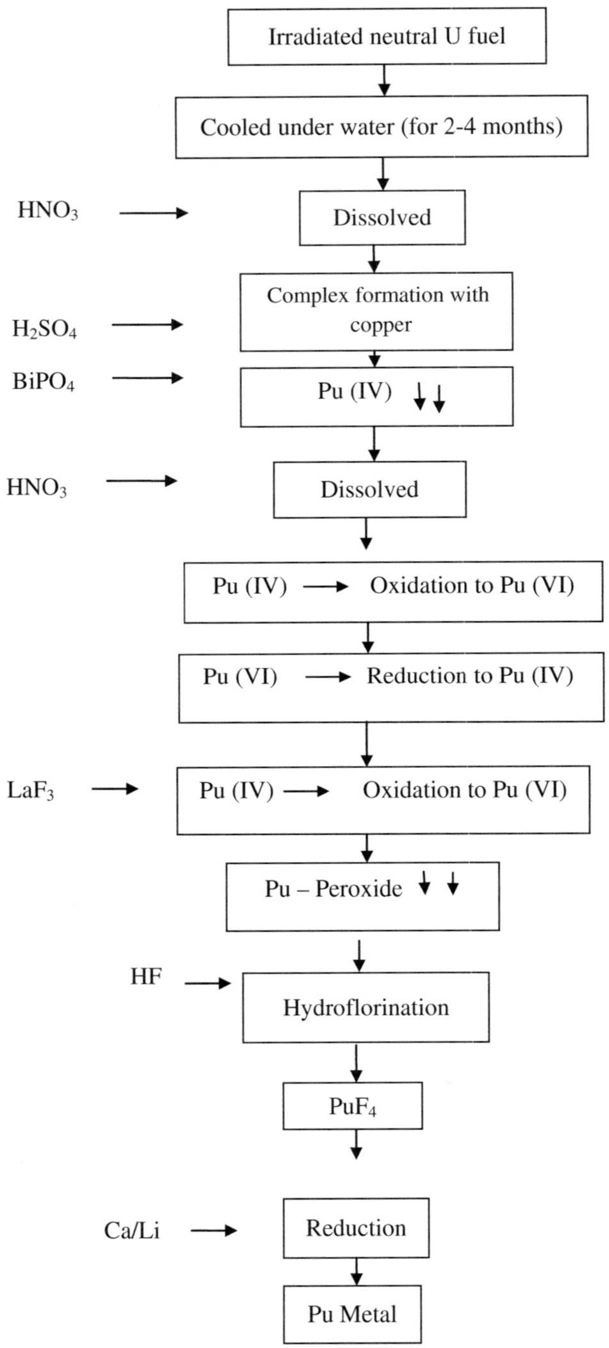

Fig. 3.1 Flow diagram for plutonium extraction

period of 2–4 months. This cooling period is required for three principal reasons: (i) to permit decay of radioactivity from fission products; (ii) to allow time for the isotope U^{239} to decay to Np^{239}; and (iii) to ensure adequate time for transformation of Np^{239} to Pu^{239} for efficient separation of the plutonium.

3.3.1.1 Precipitation

(a) Bismuth Phosphate Method ($BiPO_4$): Precipitation is the oldest of the separation methods used to recover plutonium from irradiated natural U (Fig. 3.2). $BiPO_4$ and LaF_3 are used as carrier precipitation agents. Irradiated U is dissolved in HNO_3, and H_2SO_4 is added to form a complex with uranium which prevents its subsequent precipitation. Pu (IV) is then co-precipitated with $BiPO_4$. The precipitate is again dissolved in HNO_3, Pu (IV) is oxidized to Pu (VI), and by-product precipitate of $BiPO_4$ is formed and removed. Pu (VI) remaining in solution is reduced to Pu (IV) and is again precipitate with $BiPO_4$ to repeat the decontamination cycle. Following this second cycle, the carrier is changed to LaF_3, and a similar oxidation-reduction step allows further decontamination and concentration of the plutonium. At this point, plutonium is concentrated sufficiently to allow final purification without the use of carriers, and plutonium peroxide is precipitated from the acid solution. The main disadvantage of $BiPO_4$ method is the difficulty of recovering uranium.
(b) Acetate process: Plutonium is co-precipitated with Na-Uranyl acetate which is redissolved and, after reduction of the Pu (IV), is reprecipitated, leaving Pu in solution (Fig. 3.3). Fission products are removed by repeating the cycle with less uranium carrier. The main disadvantage of this method is that the entire bulk of uranium has to be handled as a precipitate.

3.3.1.2 Organic Solvent Extraction Process

Solvent extraction techniques for separating compounds of an aqueous solution depend upon the differences in solubility of the components in an organic liquid that is essentially immiscible with water. There are two processes: (a) Purex process and (b) Redox process.

(a) Purex process: In the Purex process the feed solution of nitrates [U (IV), Pu (IV), and fission products] is extracted to separate uranium and plutonium from the bulk of the fission products (Fig. 3.4). Strength of the solution is adjusted by the addition of excess nitric acid. The organic solvent is tributyl phosphate (TBP) in a kerosene carrier, and the extraction is made in various kinds of

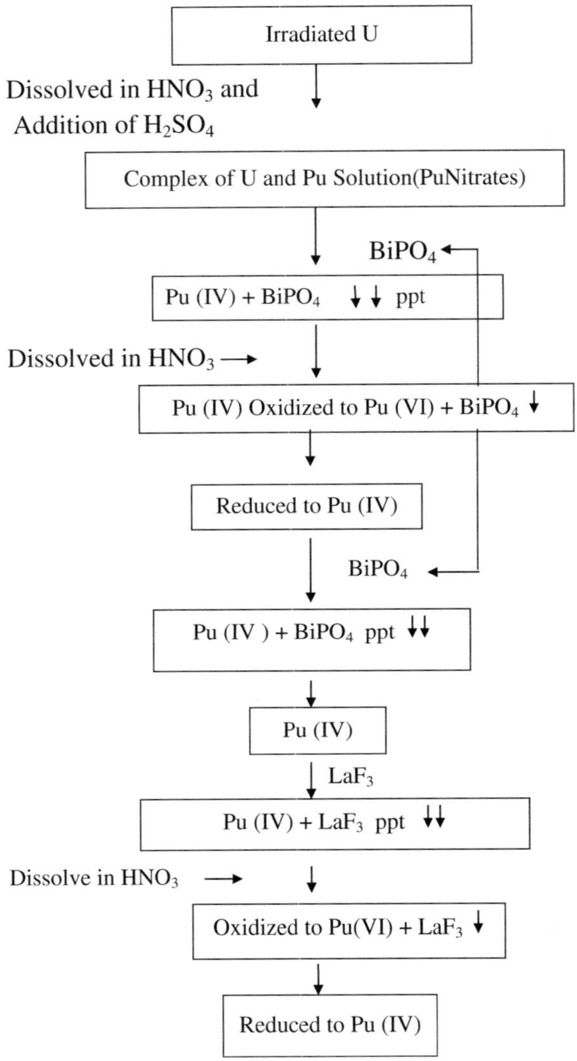

Fig. 3.2 Bismuth phosphate method

contacting streams. TBP solvent containing uranium and plutonium is contacted with dilute nitric acid in a *partitioner* which reduces the Pu (IV) to the Pu (III) state and results in the complete transfer of plutonium from the organic phase to the aqueous phase. Further purification of the plutonium is accomplished by deoxidizing it to the Pu (IV) state and re-extracting with TBP.

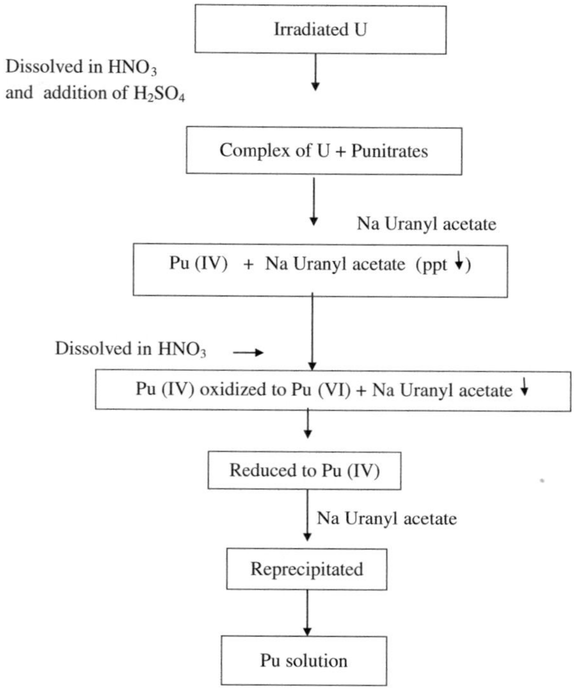

Fig. 3.3 Acetate process

(b) Redox process: Redox process is another continuous solvent extraction process that has been used successfully for several years. It performs the same functions and has the same products as the Purex process but differs in the use of organic solvent and salting agent. Redox process utilizes Hexone (methyl isobutyl ketone) as the solvent and aluminum nitrate [$Al(NO_3)_3$] as the common ion salting agent instead of the TBP and nitric acid of the Purex process. Reduction and oxidation of plutonium are accomplished with $Fe(SO_3NH_2)_2$ and $Na_2Cr_2O_7$ respectively.

3.3.1.3 Ion-Exchange Process

French metallurgists studied ion exchange methods for the isolation of plutonium; they employed a three-stage resin treatment for final purification of plutonium solution before its conversion to a dry salt and reduction to metal. First treatment is with a cation resin but the last two stages use anion resin, which compared to cation resins, appears to be inherently capable of more efficient separation of plutonium from uranium, fission products and other impurities. Ion exchange processes do not appear to be economical.

Fig. 3.4 Purex process

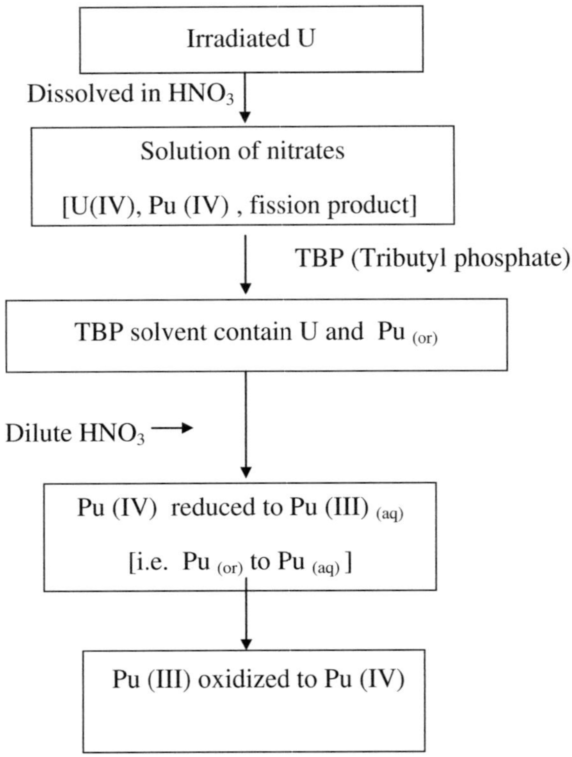

3.3.1.4 Pyrometallurgical Processes

Pyrometallurgical processes have not been developed sufficiently to have immediate application for recovering plutonium. There are two methods: (a) distillation; and (b) salt extraction.

(a) Distillation: Plutonium can be concentrated by vacuum distillation from molten uranium at the temperature range between 1500 and 1800 °C. At 1540 °C the vapour pressure of plutonium is 300 times more than that of uranium. More than 99% of plutonium has been recovered.
(b) Salt Extraction: Salt extraction of plutonium from molten uranium has been accomplished successfully by using halides such as UF_4 or $MgCl_2$. In a molten mixture of UF_4 and uranium metal (equal weight ratios of salt to metal), 90% of the plutonium goes into the salt phase. Repeated extractions should make it possible to recover practically all of the plutonium from the uranium. In this procedure the plutonium is recovered as the fluoride or chloride and later must be reduced to the metal.

3.3.2 Reduction to Plutonium Metal

The product from the organic solvent extraction process is a dilute solution of plutonium nitrate in nitric acid. The methods have been developed for conversion of plutonium salts to plutonium metal by the following steps:

(i) Preparation of a plutonium solution having a composition suitable for further processing.
(ii) Precipitation of a plutonium halide (PuF_4).
(iii) Drying of the precipitated halide or dry chemical conversion to halide.
(iv) Reduction of the plutonium halide by a more electropositive metal.

In order to avoid significant precipitation losses and to minimize the size of the processing equipment, the starting solution is concentrated to contain from 10 to 100 gm of plutonium per litre. Plutonium metal may be obtained from the concentrate by precipitation of a compound, e.g. Pu (III) fluride, which is then directly reduced to metal or by precipitating of a compound such as Pu (III) oxalate, Pu (IV) oxalate, or Pu (IV) peroxide; which is converted anhydrously to PuF_4 prior to reduction (Fig. 3.5).

Plutonium metal may be produced by reducing any plutonium halide with an alkali or alkaline earth metals:

$$PuF_4 + 2Ca \rightarrow Pu + 2CaF_2 \tag{3.5}$$

$$PuF_4 + 4Li/Ba \rightarrow Pu + 4LiF/BaF \tag{3.6}$$

$$PuCl_3 + 3/2Ca \rightarrow Pu + 3/2\ CaCl_2 \tag{3.7}$$

$$PuBr_3 + 3/2Ca \rightarrow Pu + 3/2\ CaBr_2 \tag{3.8}$$

$$PuI_3 + 3/2\,Ca \rightarrow Pu + 3/2\,CaI_2 \tag{3.9}$$

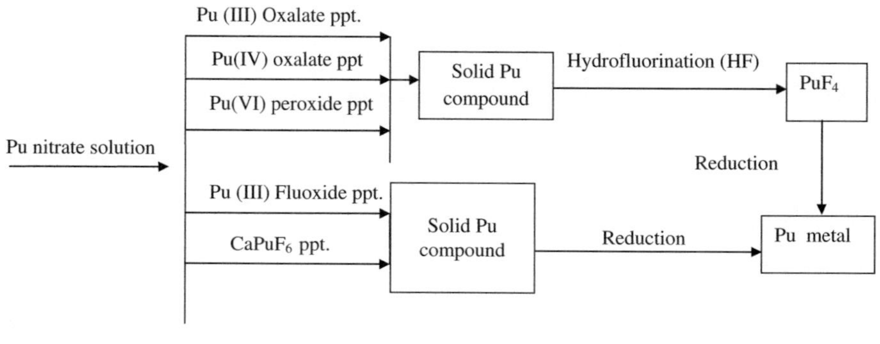

Fig. 3.5 Line diagram for plutonium metal formation from concentrate

Calcium reduction of PuF_4 is the preferred reaction, because Ca is easier to handle than Li. Reduction of PuF_4 is generally performed in an induction or resistance heated refractory lined pressure vessel at inert gas atmosphere. About 25% excess Ca is added to the PuF_4, and the reaction is initiated at about 600 °C. The addition of a booster charge containing equivalent amounts of Ca and I provides additional heat and improved slag characteristics. The optimum ratio of iodine to PuF_4 (0.1–0.5 mol I_2/mol PuF_4) is a function of the amount of charge to be reduced. The bomb reduction process is capable of yields of greater than 99%.

The purity of normal product plutonium is about 99.87 wt%. The metal may contain impurities to the extent of 500 ppm of Fe, 400 ppm of C, 200 ppm of Cr, 200 ppm of Ni, 100 ppm of Sb and 100 ppm of Si.

Neutron emission is high, and corrosion problems due to the use of HF vapour at high temperature during the fluorination step are the disadvantages of PuF_4.

Merits of $PuCl_3$ with respect to PuF_4 are: (a) neutron emission is low (1/64 neutron emission compared to PuF_4); (b) although the heat of reaction with $PuCl_3$ is only one-third of that of PuF_4, it is compensated by a slag which melts at a much lower temperature (772 °C compared to 1414 °C); and (c) during chlorination significant purification results because of high volatility of many impurity chlorides.

3.3.3 Extraction of Plutonium from Spent Fuel

In the processing of spent fuel by aqueous process the plutonium is obtained as a solution of the nitrate in HNO_3 by Purex process. From this solution Pu (IV) oxalate is precipitated by oxalic acid. By heating, Pu (IV) oxalate is converted to PuO_2. Again PuO_2 is heated with HF to obtain PuF_4.

$$PuO_2 + 4HF = PuF_4 + 2H_2O \tag{3.10}$$

PuF_4 is then reduced by excesses (25%) Ca in the presence of I (0.1–0.5 mol I_2/mol of PuF_4) at 600 °C in inert atmosphere.

$$PuF_4 + 2.3Ca + 0.3I_2 = Pu + 2CaF_2 + 0.3CaI_2 \tag{3.11}$$

Booster reaction ($Ca + I_2 = CaI_2$) supplies additional heat and CaI_2 goes to slag along with CaF_2 by lowering the melting point of the slag.

3.4 Properties

The atomic weight of plutonium is 239.06. The melting point of plutonium is 639.5 ± 2 °C, which is slightly below the melting point of aluminium (660 °C). But the boiling point of plutonium is very high (3235 ± 19 °C). It is a heavy metal, its specific gravity (19.5) is near that of uranium. Fresh surfaces have a silver-white lustre but they tarnish rapidly on exposure to air. Diamond pyramid hardness (DPH) of cast plutonium (99.5% purity) is 270. Average tensile properties of cast plutonium have been reported to be ultimate strength 43.59 kg/mm^2 (427.5 x 10^3 kPa); yield strength 26.01 kg/mm^2 (255 x 10^3 kPa), and modulus of elasticity 9.7 x 10^3 kg/mm^2 (95.2 x 10^6 kPa) respectively.

The physical properties of plutonium are dominated by the existence of the unusually large number of six allotropic modifications below this relatively low melting point. Crystal structure data for plutonium allotropies are shown in Tables 3.2 and 3.3 represents equilibrium transformation temperatures of plutonium. Difficulties in the determination of transformation temperatures of pure plutonium arise from the fact that the alpha (α) emission is of sufficient activity to result in self-heating. Between the alpha (α) and delta (δ) phases, there is a 20% decrease in density. Another peculiarity is that the delta phase possesses a negative temperature coefficient, in other words it contracts on heating.

Thermal conductivity measurements over the temperature range –163 to 140 °C have been obtained on high purity plutonium. The thermal conductivity value for α-Pu increases almost linearly with temperature, a little discontinuity at the $\alpha \rightarrow \beta$ transition temperature. All six phases of plutonium exhibit paramagnetic behaviour.

The corrosion behaviour of plutonium is in many respects similar to that of uranium. Both metals oxidize very slowly in dry air. Freshly cleaned plutonium is a bright metal. Darkening takes place rapidly in ordinary air, and attack is accelerated by moisture as uranium metal. In contrast to the attack on uranium, the attack on plutonium does not result in a thick oxide film but rather the surface becomes powdery with olive-green particles of PuO_2.

Plutonium exhibits four valence states: Pu(III), Pu(IV), Pu(V) and Pu(VI). Plutonium forms stable compounds, with all of the non-metallic elements (C, H, N, O, S, F etc.), such as the oxide PuO_2 and carbide PuC. As these exhibit no phase change and possess high melting points they are of potential as fuel element material. Plutonium resists attack by nitrogen. Very slow reaction has been

Table 3.2 Crystal structure of plutonium allotropic

Sr. no.	Allotrope	Existence Range (°C)	Crystal structures	DPH (kg/mm^2)
1	α	Below 122	Simple monoclinic ($a \neq b \neq c$, $\alpha = \gamma = 90^\circ \neq \beta$)	295
2	β	122–206	Body centered monoclinic	100[a]
3	γ	206–319	Face centered orthorhombic ($a \neq b \neq c$, $\alpha = \beta = \gamma = 90^\circ$)	110[a]
4	δ	319–451	Face centered cubic	40–70[a]
5	δ'	451–476	Body centered tetragonal ($a = b \neq c$, $\alpha = \beta = \gamma = 90^\circ$)	
6	ε	476–639.5	Body centered cubic	

[a]Hardness of high temperature phase depend on the presence of alloying element

Table 3.3 Estimated equilibrium transformation temperatures of Pu and transformation volume changes

Sr. no.	Phase transformation	Transformation temperature (°C)	Volume change (%)
1	$\alpha \rightarrow \beta$	110 ± 10	9.62
2	$\beta \rightarrow \gamma$	190 ± 10	2.67
3	$\gamma \rightarrow \delta$	310 ± 5	6.90
4	$\delta \rightarrow \delta'$	452 ± 2	−0.36
5	$\delta' \rightarrow \varepsilon$	480 ± 4	−2.16
6	$\varepsilon \rightarrow$ Liquid	639.5 ± 2	0.82

observed with pure N_2 at 250 °C, and heating of molten plutonium for several hours in N_2 at 1000 °C resulted only in a few nitride dendrites at the surface. Plutonium also reacts readily with H_2 at room temperature to form PuH_2 and PuH_3. At moderate temperature the plutonium metal reacts with halogens to form tri-halides ($PuCl_3$ etc.).

As for aqueous media, plutonium reacts chemically with slightly acid solutions, with the exception of HNO_3. The reaction in water is slower compared with the attack in acids such as HCl. Reaction with oxygen saturated water is much less aggressive than with air saturated water.

The plutonium metal is highly toxic, dust; turnings and residues are pyrophoric and subject to spontaneous combustion. Ignited, the fumes of the metal give off radioactive plutonium oxide smoke which is lethal, a fatal dose being about the size of a dust particle. The highly toxic nature of plutonium is a consequence of its radioactivity, and the permissible level for intake of plutonium by the human body is lower than for any of the other radioactive elements. Maximum permissible body burden for Pu-239 is 0.75 μ gms.

Plutonium can be fabricated into strip, rod, tube, etc., by conventional techniques. The fabrication machinery is totally enclosed in glass, operating either in a vacuum or in an inert gas atmosphere. Most operations are remotely controlled, so that the need for handling the material is reduced to a minimum.

Plutonium can be made into alloys with Co, Ni, Al, Cu, Au, Fe, Zr, Be, Hg etc. The corrosion behaviour of plutonium can be improved by alloying, that may be impacted by additions of elements which stabilize the delta phase (e.g. Al, Th, Zr). Aluminium is effective but to a lesser extent; thorium appears only slightly beneficial, e.g. Pu-3.5 at.% Al, Pu-3.5 at.% Th, Pu-7.5 at.% Zr.

3.5 Applications

At present, plutonium is used in nuclear weapons, as nuclear fuel and source of neutrons. Plutonium cannot be used for nuclear fuel in the form of the pure metal, it must be combined with other elements for reactor use. Alloys of plutonium are primarily interesting for their potential uses as fuel in nuclear reactors. Such fuels may be either of two types liquid or solid. Low melting eutectics of Pu with Fe, Co, and Ni has led to consideration of these compositions as liquid fuels. Compared to solid fuels, liquid fuels should be much less subject to radiation damage and make possible easier methods of fission product removal.

Solid reactor fuels are of two types: (a) very dilute and (b) moderately concentrate.

(i) Al-Pu alloys used as a solid fuel in nuclear reactors. 2–20 wt% (0.2–2.8 at.%) Pu contain in alloy. They are suitable for use in thermal reactors, but are unsatisfactory for fast reactors. Plutonium present as the compound $PuAl_4$ dispersed throughout a matrix of substantially pure Al, e.g. Cu-Pu (Pu_2Cu_{11} in a Cu-matrix); Fe-Pu ($PuFe_2$ in a Fe-matrix).
(ii) By moderately concentrate is meant alloys containing of the order of 5–50 at. % Pu, e.g. 15 wt% Pu in α-Th, 40 at.% Pu in Zr, Pu in α-U.

The intense alpha (α) activity of plutonium makes it important as a component of Pu-Be neutron sources, $PuBe_{13}$. The source of neutron is the spontaneous fission of Pu-240.

Pu-Al alloy has found application as driver fuel in research reactors or as booster fuel in power reaction.

$$PuF_3 + Al = Pu + AlF_3(\text{under vacuum}), \quad (3.12)$$

$$3PuO_2 + 4Al = 3Pu + 2Al_2O_3 \quad (3.13)$$

PuO_2 is preferred as the starting material because: (a) it eliminates the expensive halogenations step with its associated neutron emission problems; and (b) vacuum reduction is not necessary.

The aluminothermic reduction of the oxide is carried out in presence of excess Al and cryolite (AlF_3, 3NaF). Due to the formation of a boundary layer of Al_2O_3 the reaction is adversely slowed down. However, in the presence of cryolite, Al_2O_3 is dissolved and high yields are obtained. The amount of cryolite to be used depends to some extent on the composition of the alloy to be prepared; enough should be used so that the dissolved Al_2O_3 does not exceed 20%, otherwise the slag layer becomes too viscous and results in a low yield.

The PuO_2, aluminium and cryolite are placed in a graphite crucible and inductively heated to 1050 °C in 30 min, maintained at this temperature for 10 min. After cooling, the cryolite slag is broken away to obtain the Pu-Al alloy regular. Recovery of alloy is 92–97%.

Chapter 4
Zirconium

4.1 Introduction

Owing to zirconium's affinity for oxygen and nitrogen, which cause embrittlement, the isolation of the pure metal long eluded a satisfactory conclusion. German chemist Klaproth identified a new metal oxide in the mineral zircon in 1789. This compound was given the name zirconia and was recognized to contain a new element. In 1824 Berzelius produced first impure metallic zirconium (Zr) by the reduction of K_2ZrF_6 with potassium metal. But it was not until some 80 years later that a malleable zirconium metal was produced.

A notable feature of zirconium is its resistance to acids and alkalis. Zirconium metal is difficult to handle in the molten condition, since it either reduces or dissolves and becomes contaminated with most refractories; that metal becomes too brittle for any use.

Zirconium (Zr) based alloys find application in nuclear power reactors as the cladding material because of their high mechanical strength, corrosion-resistant properties and low capture cross-section for thermal neutron (0.18 barn). Hafnium (Hf), which co-occurs along with zirconium in the ore, and has a high absorption cross-section for thermal neutrons (105 barns). Nuclear grade zirconium contains less than 100 ppm of hafnium.

4.2 Sources

The chief source of zirconium minerals is the beach sands of Queensland in Australia, where the silicate mineral zircon occurs associated with rutile. Zircon is also present in the monazite sands of India. The occurrence of zirconium in the Earth's crust is greater than the total percentage of copper, lead and zinc.

S.K. Dutta and D.R. Lodhari, *Extraction of Nuclear and Non-ferrous Metals*,
Topics in Mining, Metallurgy and Materials Engineering,
https://doi.org/10.1007/978-981-10-5172-2_4

The major source of zirconium is zircon sand [Zr(Hf)SiO_4], which is found in Australia, USA and India, with the sea-beach sand. Baddeleyite ore (ZrO_2) is found in Brazil. This ore, when it contains some zircon, is often called zirkite.

Sea-beach sands are readily concentrated by gravity methods; the use of shaking tables, which are simpler and use less power. Final cleaning of the concentrate is accomplished with magnetic and electrostatic separators.

4.3 Exreaction of Zirconium

4.3.1 Separation of Zirconium and Hafnium

Zircon sand contains 0.5–2.0% Hf, and the hafnium has chemical properties so similar to zirconium that no separation is achieved in the usual chemical processes. Reactor grade zirconium is limited to a maximum of 0.01% Hf, because the later metal has such a high thermal neutron absorption cross-section. Zirconium has a cross-section of 0.18 barns, whereas hafnium has a cross-section of 115 barns which is more than 500 times that of zirconium. For this reason hafnium is used as control rods in reactors.

Many methods have been developed for the separation of zirconium and hafnium (flow sheet for the extraction of Zr and Hf metals is shown in Fig. 4.1), such as fractional distillation, ion exchange, fractional crystallization, liquid-liquid extraction etc. The last named procedure is the most widely used. In one process, crude zirconyl chloride is brought into contact with a counter-current stream of thiocyanate-rich methyl isobutyl ketone.

(a) Solvent Extraction: The solvent extracts of hafnium and the zirconyl chloride are mixed with H_2SO_4 and NH_4OH to produce zirconium hydroxide, which is subsequently heated to produce pure zirconium oxide (ZrO_2). Other solvents and procedures of precipitation are used. Increased interest has been shown in a process wherein $ZrCl_4$ is selectively reduced to $ZrCl_3$; the $HfCl_4$ is then sublimed and a separation is achieved.
(b) TBP- Nitric Acid process: Zircon sand is fused with NaOH at 600 °C to form zirconium-hafnium hydroxide which is leached with water to remove the solute sodium silicate.

$$Zr(Hf)SiO_4 + 4NaOH = Zr(Hf)(OH)_4 + Na_4SiO_4 \quad (4.1)$$

The cake is dissolved in nitric acid (50–60%) and form zirconium-hafnium nitrate solution.

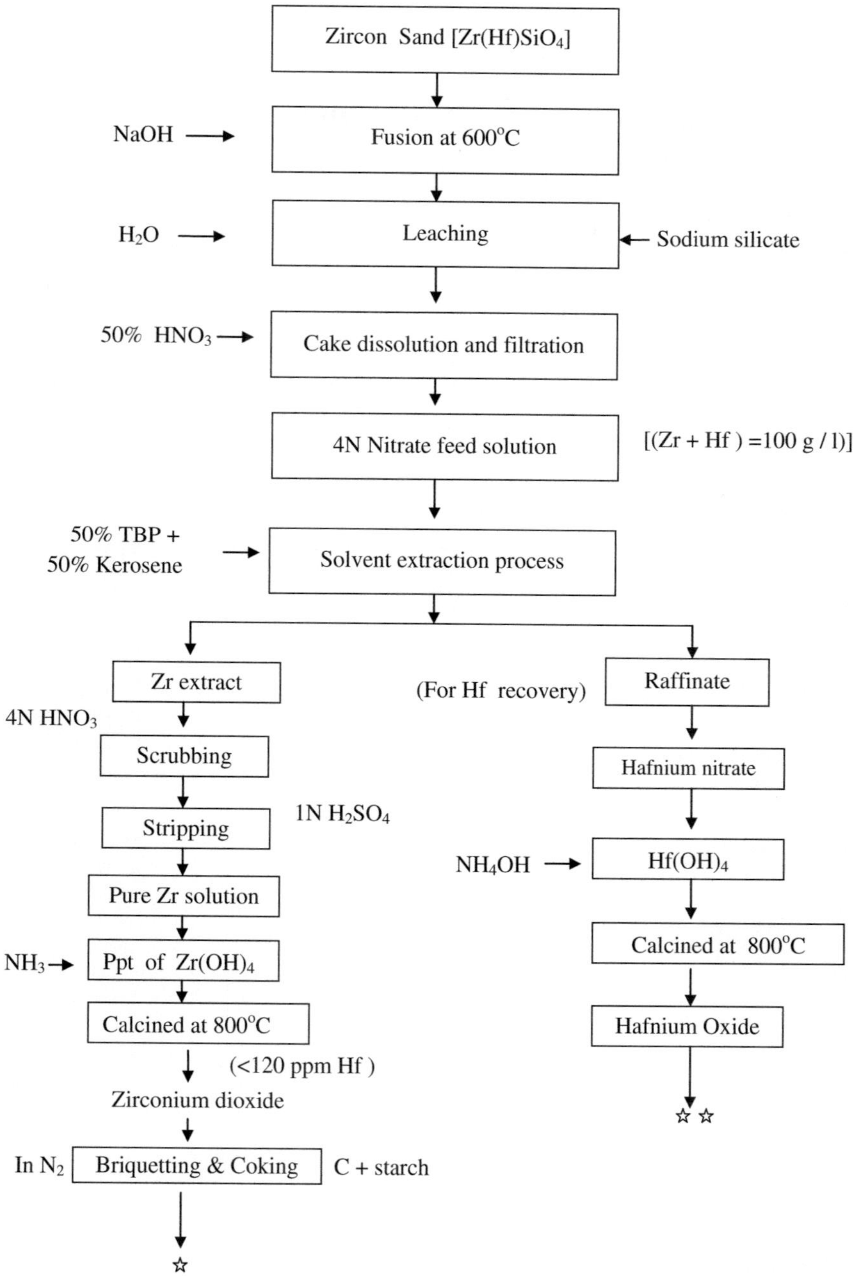

Fig. 4.1 Flow sheet for the extraction of zirconium and hafnium metals

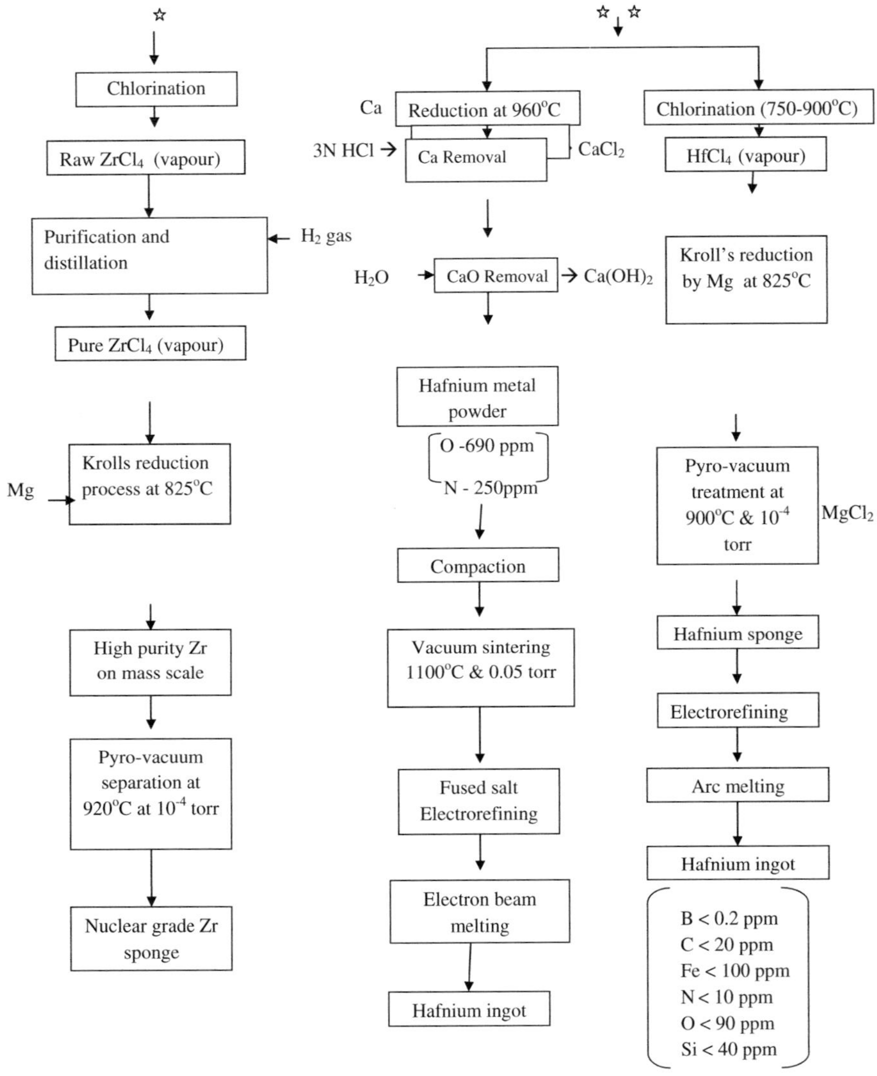

Fig. 4.1 (continued)

$$Zr(Hf)(OH)_4 + 4HNO_3 = Zr(Hf)(NO_3)_4 + 4H_2O \quad (4.2)$$

Now zirconium and hafnium are separated by solvent extraction method by using 50% TBP in kerosene. So zirconium is collected as an extract in organic phase and hafnium is separated as a raffinate in aqueous phase. The scrubbing of zirconium by concentrate (4 N) HNO_3 and stripping with dilute (1 N) H_2SO_4 brings zirconium back into the aqueous phase (pure zirconium solution).

4.3.2 *Preparation of Zirconium Oxide*

After getting the pure zirconium solution (nitrate), zirconium hydroxide is then precipitated by ammonia.

$$Zr(NO_3)_4 + NH_3 \rightarrow Zr(OH)_4 + HNO_3 \quad (4.3)$$

This zirconium hydroxide is calcined at 800 °C, to form ZrO_2.

$$Zr(OH)_4 = ZrO_2 + 2H_2O \quad (4.4)$$

4.3.3 *Production of Zirconium Tetrachloride*

Direct metallothermic reduction of zirconium oxide is not attractive for the production of ductile zirconium, since it is difficult to eliminate residual oxygen from the resultant powder zirconium. So the conversion of zirconium oxide to the metal on a commercial scale thus involves chlorination of zirconium oxide to obtain zirconium tetrachloride. Chlorination of zirconium oxide is carried out by reacting with chlorine and carbon mixture in a static chlorinator reactor.

Firstly, zirconium dioxide and carbon mixture (with starch as a binder) is briquetted in nitrogen atmosphere to form porous briquettes and heated in a silica lined furnace. This zirconium dioxide and carbon briquette possesses sufficient electrical conductivity, which improves with temperature, and with appropriate voltage and current control; the briquettes are slowly heated to the reaction temperature of 750 to 900 °C.

$$ZrO_2 + 2C + 2Cl_2 = ZrCl_4 + 2CO \quad (4.5)$$

$ZrCl_4$ vapour evolving from the chlorination is condensed as dense crystals in a double walled inconel condenser at 150–200 °C. The raw chloride is then purified by sublimation at 350 °C (boiling point of $ZrCl_4$ is 334 °C) in hydrogen atmosphere.

4.3.4 *Reduction of $ZrCl_4$ by Mg or Na*

Since zirconium is highly corrosive in the molten form and extremely so in the powder form, it is essential to produce it in the partially sintered condition as obtained in the Kroll process.

4.3.4.1 Kroll Process (by Mg)

The reduction of $ZrCl_4$ with Mg produces a high purity zirconium metal. This method was developed by scientist W.J. Kroll and his co-workers at the United States Bureau of Mines, hence this method is known as Kroll process. The Kroll process comprises a specially designed three-zone furnace with provision to maintain temperature in each zone. The unique feature of the Kroll process is the heating of the zirconium chloride so that only the vapours come in contact with the magnesium which is the reducing agent.

Zirconium chloride ($ZrCl_4$ sublimate at 334 °C) and magnesium are loaded in separate containers in the retort. The chloride is purified by eliminating volatile impurities at 300 °C and temperature of Mg is raised to 850 °C. The chloride is distilled under control to react at the surface of molten Mg (melting point 651 °C). For optimum reduction rates the process has to be controlled in the temperature range 800–875 °C.

$$ZrCl_4 + 2Mg = Zr + 2MgCl_2 \tag{4.6}$$

At the end of the reduction the crucible contains a sponge like network of zirconium crystals covered with $MgCl_2$ and unreacted Mg. Mg and $MgCl_2$ are then removed by pyro-vacuum separation at 920 °C and 10^{-4} torr pressure in a separate distillation retort to get nuclear grade zirconium sponge which contain impurities (B < 0.5, C − 400, N − 85, O − 865, Hf − 120 ppm).

4.3.4.2 Modified Kroll Process (by Na)

Reduction by sodium has certain advantages in that its low melting point (97.5 °C) makes it possible to handle it in liquid form; this facilitates purification, handling, and feeding to the reactor. The reaction products zirconium and NaCl can be leached with water or weak acids.

$$ZrCl_4 + 4Na = Zr + 4NaCl \tag{4.7}$$

The use of sodium has certain disadvantages: (a) nearly twice as much mono-valent sodium is used per kg of zirconium as compared to di-valent magnesium; (b) there is a greater quantity of by-product NaCl is to be disposed of; and (c) low boiling point of sodium (883 °C) and high melting point of NaCl (801 °C) also created some problems.

4.3.4.3 Bimetal Reduction

The use of a small percentage of sodium along with magnesium has been found to form a purer product. This is attributable to the guttering ability of sodium, reduced iron contamination (iron coming from the crucible of the reactor), better flexing ability of NaCl with impurity chlorides (such as $FeCl_3$, $AlCl_3$) and better removal of $MgCl_2 - NaCl$ slag (which is more fluid than $MgCl_2$ alone).

4.4 Properties

Zirconium is found to be steel-grey in colour, not quite so heavy as iron, with tensile properties similar to mild steel. Zirconium has an atomic weight of 91.22. The specific gravity of zirconium is 6.45, its melting point and boiling point are 1852 ± 10 °C and 3580 – 3700 °C respectively. Zirconium has an allotropic transition HCP to BCC ($\alpha \rightarrow \beta$) at 862 ± 5 °C. Transition from $\alpha \rightarrow \beta$ will be increased with present of O, N, Sn and Al; and decreased with Ta, Cb, Th and U. Zirconium, if sufficiently pure, is soft and ductile. Hardness of zirconium is increased with increasing oxygen.

Zirconium reacts readily with oxygen, the rate of reaction varying with particle size, temperature presence of impurities, and the nature of the surface. It was found that large amounts of oxygen, up to 60 atomic%, can be dissolved in solid solution in zirconium without showing any ZrO_2 structure. The equilibrium vapour pressure of saturated solutions of oxygen in zirconium is so low that there is no possibility of removing oxygen from zirconium by vacuum treatment. Of the oxygen content in zirconium, 50–90% can be removed by heating in molten Ca or in Ca vapour.

The reaction of nitrogen gas and zirconium occurs very slowly at 400 °C, but increases rapidly at 800 °C and above; the reaction is insensitive to pressure. Amount up to 20 atomic% form solid solution; above this ZrN is formed. Nitrogen gas cannot be removed by vacuum treatment or with calcium.

Hydrogen gas is absorbed rapidly by zirconium at temperatures from 300 to 1000 °C. In contrast to oxygen and nitrogen gases, hydrogen gas can be completely removed by prolonged heating in a vacuum above 1000 °C. The absorption of hydrogen gas embrittles zirconium to the extent that it can be crushed, thus providing one method for producing zirconium powder.

Zirconium reacts rapidly with CO_2 gas above 800 °C; zirconium also reacts with CO gas and H_2O vapour above 1000 °C. Zirconium oxide (ZrO_2) is formed by all these gases, and at high temperature CO and CO_2 gases also lead to zirconium carbide formation. Zirconium reacts readily with all of the halogens in the range of 200–400 °C to form volatile halides. Zirconium also reacts with S, C, Si, P, B and Al at elevated temperatures.

A notable feature of zirconium is its resistance to acids, hot nitric and sulphuric acids having little effect; it is insoluble in hydrochloric acid in concentrations up to 20%. Resistance to alkalis is also a prominent characteristic.

Reactor grade unalloyed zirconium is sufficiently pure to withstand the corrosive attack of water indefinitely at temperature up to 260 °C. Above this temperature, the behaviour becomes erratic. For this reason, the zircaloy alloys were developed for use in pressurized water reactors. The principal alloy in use is zircaloy-2, which consists of 1.5 wt% Sn, 0.12 wt% Fe, 0.1 wt% Cr and 0.05 wt% Ni in Zr. The presence of Fe, Cr and Ni retard accelerated corrosion and film failure. The corrosion behaviour of Zr-alloys in high temperature water and steam is very sensitive to the cleanliness of the metal surface. Although the material is readily machinable, it is easily overheated, and the surface layer becomes contaminated by nitrogen.

Complete solid solubility is found in the Zr-Hf and Zr-Ti systems. An eutectic and an eutectoid are obtained with H, W, V, Ag, Be, Co, Cr, Cu, Fe, Mn, Mo and Ni. The tensile strength of Zr is as low as 17.575 kg/mm^2. Addition of Mo, Sn and Nb to Zr improves tensile strength (140.6 kg/mm^2) and results in a fair amount of ductility (6–10%). Unalloyed uranium was extrusion clad with two of these alloys Zr-5 wt% Mo-2 wt% Sn, and Zr-5 wt% Nb-2 wt% Sn. Al may replace Sn in the later alloy, and strength may be maintained.

Zirconium is inert to pure liquid sodium, and its reaction with liquid metal coolant is confined to guttering the oxygen impurity from sodium. An adherent oxide is formed on the zirconium, and oxygen diffuses into the metal. The rates for both oxidation and diffusion of oxygen in zirconium are strongly temperature-dependent.

Zirconium ingot can be heated to 850 °C in ordinary furnace atmosphere and rolled or forged. Seamless tubing is made by extruding billets coated with glass or encased in copper. Salt bath heating permits extrusion temperature up to 1000 °C. Zirconium must be welded in an inert atmosphere to prevent harmful contamination by atmospheric gases.

4.5 Applications

The chief consumption of zirconium metal has been for nuclear energy applications. These result from its excellent corrosion resistance combined with a low absorption cross-section for neutrons. The cross-section for zirconium is 0.18 barn (1 barn = 10^{-24} sq cm/nucleus) as compared with 0.22 barn for aluminium, and 3 barns for stainless steel, and 120 barns for hafnium. Zirconium can be used in the core of a nuclear reactor as a cladding material for the fuel elements, as an alloying metal with uranium.

The general commercial application of zirconium depends mainly on its corrosion resistance. It combines outstanding resistance to both mineral and organic acids as well as to strong alkaline solutions. Pumps, valves, heat exchangers, filters, piping, and other components are produced by commercial grade zirconium. These components are used in chemical industry. Other uses, such as in vacuum tubes, in photoflash bulbs, and for surgical use, make it a popular metal.

High strength zirconium alloys are developed for two reasons:

1. In natural uranium fueled heavy water moderated reactors, neutron economy is of prime consideration. If a reduction of cladding thickness could be achieved by using a high strength zirconium alloy to replace zircaloy-2, which is only a

little stronger than zirconium, then a decrease in parasitic neutron capture by the cladding would be achieved.

2. High strength claddings may be used to constrain the swelling associated with irradiation of uranium alloy fuels and thereby increase the allowable fuel burn-up.

For chemical industry use, zirconium combines extreme resistance to corrosion in both alkaline and acids exposures. Its low corrosion rates in acids, dry chlorine (below 200 °C), alkalis, sea water, halide salts, and sodium hypochlorite are outstanding. These factors, combined with its fine mechanical properties, make its use attractive in chemical equipment.

Chapter 5
Hafnium

5.1 Introduction

Hafnium (Hf) occurs in nature in small amounts associated with zirconium in all the zirconium minerals. The hafnium content of the Earth's crust has been estimated to be greater than that of mercury, niobium, and tantalum or silver. Although hafnium was not discovered until 1923, claims of evidence of the concentration of this element were made many years earlier. The evidence was considered inconclusive by Bohr and by Coster and von Hevesy, who were convinced that element 72 was not one of the rare earths but that it was quadrivalent and a homologue of zirconium and thorium. Coster and von Hevesy reasoned that element 72 was more likely to occur with zirconium than with the rare earths. Coster and von Hevesy announced the discovery of element 72, proposing the name *hafnium* (from *Hafnia*, Latin for Copenhagen) in honour of the city where the discovery was made.

There has been considerable interest in the use of hafnium (Hf) metal as a control rod material in nuclear reactors. This use has been made possible in large measure by the somewhat greater availability of the metal as a by-product of the increased production of reactor grade zirconium. Hafnium excels in this application because of its excellent hot water corrosion resistance, good ductility, machinability and high thermal neutron absorption cross-section (105 barns).

5.2 Sources

Hafnium occurs in nature in small to moderate amounts associated with zirconium in all types of zirconium-bearing minerals. The ratio of hafnium to zirconium in the Earth's crust has been estimated at about 0.02. The principal commercial sources of hafnium are the minerals zircon [$Zr(Hf)SiO_4$] and baddeleyite (ZrO_2), which are processed primarily for their zirconium content. These minerals contain 0.5–2.0% Hf.

S.K. Dutta and D.R. Lodhari, *Extraction of Nuclear and Non-ferrous Metals*,
Topics in Mining, Metallurgy and Materials Engineering,
https://doi.org/10.1007/978-981-10-5172-2_5

5.3 Extraction of Hafnium

5.3.1 Separation of Zirconium and Hafnium

Hafnium and zirconium are always extracted together and since the separation of hafnium from zirconium is an important step in the production of reactor grade zirconium. The flow sheet for the extraction of Zr and Hf metals is shown in Fig. 4.1. TBP-nitric acid process (as described in Sect. 4.3.1 in extraction of Zr). The raw material for production of metal is hafnium oxide obtained as a by-product in the production of hafnium free zirconium.

5.3.2 Preparation of HfO_2

Hafnium nitrate in aqueous phase is achieved from solvent extraction process, then $Hf(OH)_4$ is precipitated by NH_4OH.

$$Hf(NO_3)_4 + 4NH_4OH = Hf(OH)_4 + 4NH_4NO_3 \tag{5.1}$$

This $Hf(OH)_4$ is calcinated at 800 °C, to form HfO_2.

$$Hf(OH)_4 = HfO_2 + 2H_2O \tag{5.2}$$

5.3.3 Production of Hafnium Metal

Hafnium metal is prepared by the same methods which are used into the production of zirconium: (a) Kroll process; (b) Modified Kroll process (using Na as the reducing agent); (c) reduction of HfO_2 by Si/Al/Ca.

5.3.3.1 Kroll Process

Chlorination of HfO_2 is done by chlorine gas in presence of carbon at 750–900 °C.

$$HfO_2 + 2C + 2Cl_2 = HfCl_4 + 2CO \tag{5.3}$$

Sublimation point of $HfCl_4$ is 316 °C, so resublimed at 325 °C of $HfCl_4$ in an inert atmosphere for purification.

$HfCl_4$ and Mg (40% excess) are taken in Kroll reduction furnace at the temperature of 825 °C for reduction.

$$HfCl_4 + 2Mg = Hf + 2MgCl_2 \qquad (5.4)$$

The produced hafnium metal is covered by the excess magnesium and molten $MgCl_2$. Then by pyro-vacuum treatment at 900 °C and 10^{-4} torr pressure separate $MgCl_2$ and excess Mg to get hafnium metal sponge (which contain 1000 ppm of O). The sponge metal is pressed into bars and cast into ingots, using the consumable electrode arc technique.

De-Boer or hot wire process or iodide refining: This process developed by Van Arkel and De Boer. The product of the hot wire process is a rod, or crystal bar of compact, ductile hafnium. The crude metal is maintained at 600 °C and the filament at 1600 °C. The iodine process must be considered a refining method rather than a basic reduction step.

5.3.3.2 Reduction of $HfCl_4$ by Na

The use of Na to replace Mg for the large-scale reduction of the tetrachlorides of Zr and Hf are developed. The main advantage of Na reduction method is its lower cost.

$$HfCl_4 + 4Na = Hf + 4NaCl \qquad (5.5)$$

The reduction step is followed by leaching of the product with water to remove salt and excess Na from the hafnium metal sponge.

5.3.3.3 Reduction of HfO_2 by Si/Al

(a) Attempts have been made at the metallothermic reduction of the oxides of hafnium in a vacuum, using Si, $CaSi_2$ and Al as reducing agents. During the reduction in a high vacuum (use of an electron beam furnace) mainly gaseous suboxides like SiO and Al_2O are formed. Besides these the volatile suboxide of Hf(HfO) is also formed.

The reduction of hafnium oxide by Si proceeds at temperatures between 2200 and 3200 °C according to:

$$HfO_2(s) + 2Si(l) = Hf(l) + 2SiO(g) \qquad (5.6)$$

A slight deficiency in silicon should be maintained in the starting mixture in order to avoid residual silicon content in the final product which is difficult to remove. Excess oxygen is mainly removed as gaseous HfO. The composition of the evaporated oxide changes with increasing reaction time, and also the losses in metal owing to evaporation increase.

The metal obtained is workable only within limits. Reduction by $CaSi_2$, which appears to be more favourable, results in a more thorough deoxidation, but the nitrogen content of the $CaSi_2$ (0.04%) is entirely transferred into the hafnium, which will contain 0.15% N.

$$3HfO_2(s) + 2CaSi_2(l) = 3Hf(l) + 2CaO(s) + 4SiO(g) \tag{5.7}$$

(b) **Ca Reduction of HfO_2 or BARC process**: This process was developed at the Metallurgy Division, Bhabha Atomic Research Centre (BARC), Trombay, India. HfO_2 is reduced by Ca at 960 °C:

$$HfO_2 + 2Ca = Hf + 2CaO \tag{5.8}$$

The excess Ca is removed by (N) HCl as $CaCl_2$ and by-product CaO is removed by water.

$$Ca + 2HCl = CaCl_2 + H_2 \tag{5.9}$$

$$CaO + H_2O = Ca(OH)_2 \tag{5.10}$$

After removal of excess Ca and CaO, Hf metal powder is washed by acetone and dried in vacuum. This dried hafnium powder (which contain 6900 ppm O and 250 ppm N) is compacting and vacuum sintering at 1100 °C, and 0.05 torr pressure. Now sintered hafnium metal is refined by fused salt electro-refining process; after that impurities contain 10% Zr, 126 ppm O and 10 ppm N. Hence, hafnium metal is again remelted by electro beam melting for refining to get hafnium ingot.

5.4 Properties

Metallic hafnium has a bright lustre. It is harder and less easily workable than zirconium. Like zirconium, it crystallizes in the close packed hexagonal system, and in many other physical properties is similar to zirconium. Properties of hafnium are very much dependent on the impurity content in hafnium. The atomic weight of hafnium is 178.5. The specific gravity of arc melted crystal hafnium bar (which contain less than 100 ppm Zr) is 13.29. Melting point of that hafnium bar is 2150 °C, and melting point of crystal bar hafnium (which contains 80 ppm Zr) is 2222 ± 30 °C. Boiling point of hafnium is 4600 °C. It can be hot and cold worked in much the same manner as zirconium.

Allotropic transition temperature from α (hexagonal close packed) to β (body centred cubic) is vary from as low as 1310 ± 10 °C to as high as 1950 ± 10 °C (which is close to the melting point of hafnium).

Chemical properties of hafnium are much closed to zirconium. Unlike zirconium, however, it is less sensitive to pick-up of nitrogen and oxygen. Hafnium is oxidized under 1 atm of O_2 pressure in the range of 350 to 1200 °C. The resistance of hafnium to air oxidation at elevated temperature is fairly good and somewhat greater than that of zirconium. Hafnium has good resistance to high temperature, high pressure water up to 399 °C and 2.46 kg/mm^2.

5.5 Applications

Hafnium has few commercial uses because of its limited supply and high price, which have been due to the difficulty in separating hafnium from zirconium. Hafnium's main use has been for making rods in controlling the energy level in nuclear reactors and as structural material in water-cooled nuclear reactors where its corrosive resistance has proved superior to that of zirconium. In addition to having an adequate absorption cross-section for thermal neutrons, hafnium has excellent mechanical properties and is extremely corrosion resistance.

Hafnium has been used as a filament in incandescent lights, as a cathode in X-ray tubes, and as an electrode (alloyed with W or Mo) in high pressure discharge tubes. Powder hafnium has been used with Ba or Sr oxide in cathodes for high vacuum discharge tubes.

Hf–Ti alloys which are free from O, N, C and Si may be used as gas filled devices such as lamps, radio-tubes, and television tubes. Hafnium has been used also in rectifiers. Alloys of hafnium with Mn, Cr, Fe, Co, Ni, Cu and Ag have been prepared. An alloy containing 0.5% Hf with about 80% Ni and 19.5% Cr is used for electrical resistance heating elements.

The hafnium oxide is a valuable insulation for very high temperature and also serves as a raw material for high temperature ceramics. The carbide in conjunction with tantalum carbide provides a compound which possesses the highest recorded melting point (4000 °C) for ceramics.

Chapter 6
Thorium

6.1 Introduction

The discovery of thorium (Th) is attributed to Berzelius in 1828. The name thorium was derived from *Thor*, Scandinavian god of war. Interest in the metal and its oxide lagged until about 1884. The interest in thorium for nuclear energy application is increased owing to its abundance in the Earth's crust 0.0012% thorium as compared to 0.0004% uranium. Thorium is a major potential source for increasing the world's supply of nuclear fuel. It is a fertile material, for producing U^{233}, in a nuclear reaction similar to the conversion of U^{238} to Pu^{239}. Th^{232} captures a neutron and subsequently emits two β particles to become U^{233}, a long-lived fissionable isotope.

$$_{90}Th^{232} + {}_{0}n^{1} \rightarrow {}_{90}Th^{233} + \gamma \tag{6.1}$$

$$_{90}Th^{233} \rightarrow {}_{91}Pa^{233} + {}_{-1}\beta^{0} \tag{6.2}$$

$$_{91}Pa^{233} \rightarrow {}_{92}U^{233} + {}_{-1}\beta^{0} \tag{6.3}$$

Thorium, like uranium, is a radioactive element in nature and emits α particle. The metal has a high density, low strength, and poor corrosion resistance in water. These properties, together with its high cost, have prevented its use as an engineering material. The use of thorium metal instead would offer the advantages of case of fabrication, and good thermal conductivity. Thorium metal has a relatively high melting point (1750 °C) and has good dimensional stability under irradiation.

S.K. Dutta and D.R. Lodhari, *Extraction of Nuclear and Non-ferrous Metals*,
Topics in Mining, Metallurgy and Materials Engineering,
https://doi.org/10.1007/978-981-10-5172-2_6

6.2 Sources

Thorium occurs in a large number of minerals associated with uranium and the rare earths metals. Those which contain a high amount of the element, such as thorianite (ThO_2 + UO_2) and thorite ($ThSiO_4$) are scarce.

The major source of the thorium metal is monazite. This mineral occurs associated with silica and other minerals in the so-called monazite sands. The largest deposits of these sands are in India and Brazil. Other countries are Australia, Ceylon, South Africa, Canada and the USA.

The major sources of monazite in India are found in Kerala (at the sea-coast of Travancore) and Tamil Nadu; monazite is also found on Bihar. The *black sands* of the Kerala sea-coast contain ilmenite ($FeTiO_3$), garnet, monazite, quartz, rutile and other minerals. They get their name from the black colour of the ilmenite. The monazite content of these sands is 4–5%. First separation done by nature is carried further by man with the help of magnetic or electrostatic separators, a concentration of 97–99% of monazite is obtained. Monazite contains thoria (ThO_2), uranium oxide (U_3O_8), phosphate and rare earths metals oxides. Monazite of the Kerala sea-coast contains 8–9% ThO_2, 0.3–0.4% U_3O_8, 59% RE_2O_3 (where RE are cerium, Ce; lanthanum, La: holmium, Ho; thulium, Tu etc.).

6.3 Extraction of Thorium

Extraction of thorium metal from monazite sand is shown in flow-diagram (Fig. 6.1). Thorium oxide is present in monazite in rather lower concentration (8–9%) than cerium (Ce), lanthanum (La) and yttrium (Y) rare earths metal oxides (59%). This created technical difficulties for separation of thorium from rare earths metals. There are, in general, two chemical processes commonly used for separation: (a) sulphuric acid method; (b) caustic soda method.

6.3.1 Separation of Thorium Compound from Monazite

6.3.1.1 Sulphuric Acid Method

Separation of thorium compound from monazite by sulphuric acid method has been in commercial use for many decades. The monazite sands are decomposed chemically in an excess of concentrated (93%) H_2SO_4 at a temperature of 210 °C for a period of 1 to 5 h, depending on the sand particle size. The pasty mass which is produced is cooled and cold water is added to dissolve the thorium, uranium, and rare earth metals (RE). Thus, monazite sulphate solution is formed and separated from the insolubles, primarily consisting of silica.

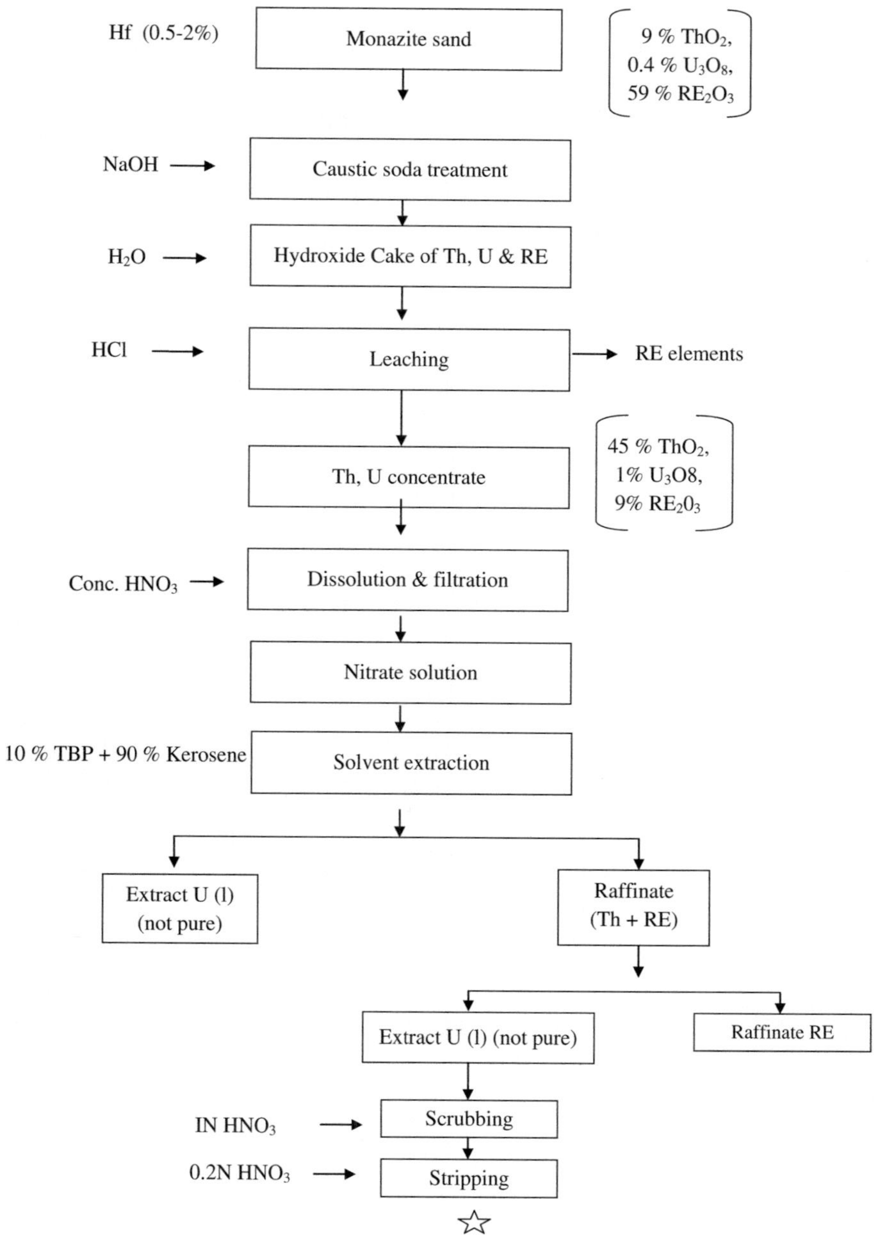

Fig. 6.1 Flow sheet for the extraction of thorium metal

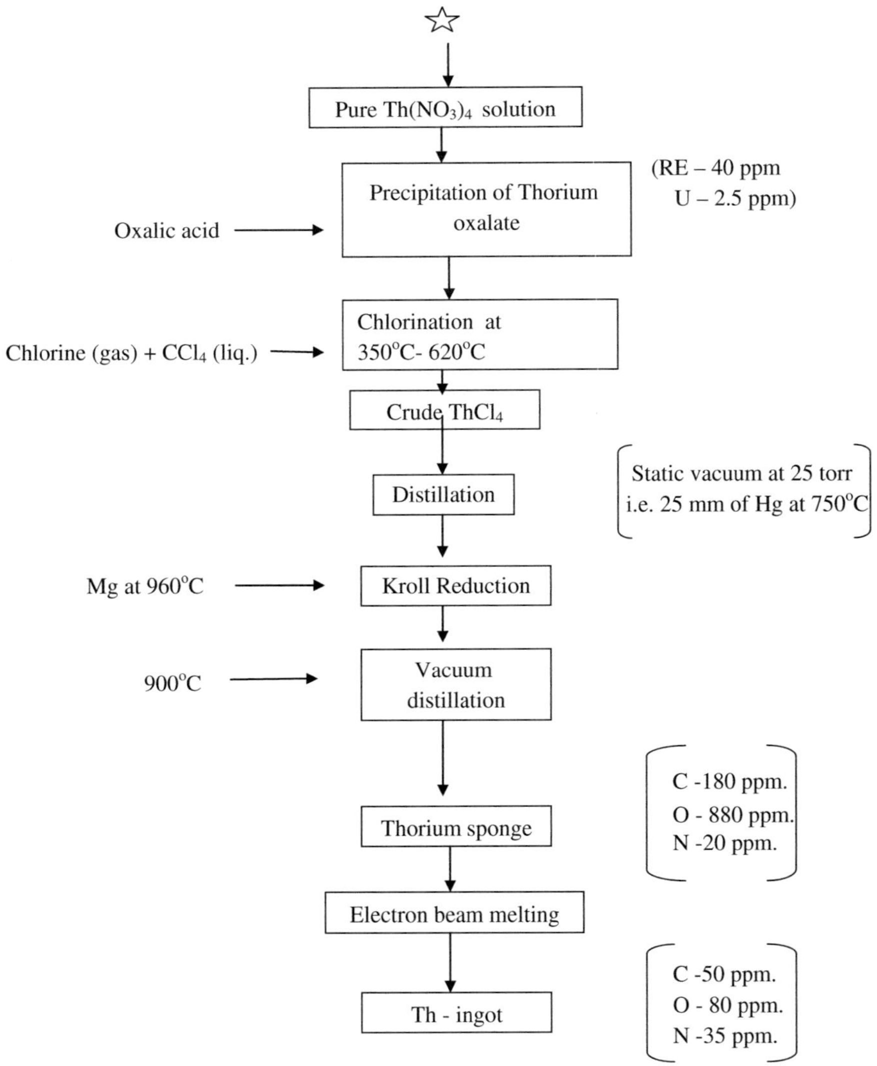

Fig. 6.1 (continued)

$$ThO_2 + 2H_2SO_4 = Th(SO_4)_2 + 2H_2O \tag{6.4}$$

$$U_3O_8 + 7H_2SO_4 = 3UO_2(SO_4)_2 + 6H_2O + H_2S \tag{6.5}$$

$$RE_2O_3 + 3H_2SO_4 = RE_2(SO_4)_3 + 3H_2O \tag{6.6}$$

Then that is treated with ammonium hydroxide to get a p_H of 1.0 for the solution. Some 99% of the thorium and about 5% of the rare earths metals are precipitated from the solution.

$$Th(SO_4)_2 + 4NH_4OH = Th(OH)_4 + 2(NH_4)_2SO_4 \tag{6.7}$$

$$RE_2(SO_4)_3 + 6NH_4OH = 2RE(OH)_3 + 3(NH_4)_2SO_4 \tag{6.8}$$

Because of the much larger quantities of rare earths present originally in the ore, the precipitates are filtered off and treated with nitric acid to redissolve the thorium, which is then extracted selectively by solvent extraction method with TBP (tributyl phosphate).

$$Th(OH)_4 + 4HNO_3 = Th(NO_3)_4 + 4H_2O \tag{6.9}$$

The sulphuric acid method has the disadvantage of allowing sulphate and phosphate ions to remain with the thorium; those ions interfere with the subsequent purification process.

6.3.1.2 Sodium Hydroxide Method

Finely ground (−325 mesh) monazite sands are chemically decomposed with 45% concentrated sodium hydroxide (NaOH) solution at 138 °C for about 3 h to obtain precipitations of Th, RE and U hydroxides.

$$ThO_2 + 4NaOH = Th(OH)_4 + 2Na_2O \tag{6.10}$$

$$U_3O_8 + 12NaOH = 3U(OH)_4 + 6Na_2O \tag{6.11}$$

$$RE_2O_3 + 6NaOH = 2RE(OH)_3 + 3Na_2O \tag{6.12}$$

The residue is concentrated by evaporation to form a cake, which is dissolved in 37% HCl at 80 °C. HCl dissolves the bulk of the rare earths metals and partially neutralized with NaOH to p_H 5.8. Thus all the thorium, uranium and only a small percentage of some rare earths metals are precipitated.

$$Th(OH)_4 + 4HCl = ThCl_4 + 4H_2O \tag{6.13}$$

$$U(OH)_4 + 4HCl = UCl_4 + 4H_2O \tag{6.14}$$

$$RE(OH)_3 + 3HCl = RECl_3 + 3H_2O \tag{6.15}$$

$$ThCl_4 + 4NaOH = Th(OH)_4 + 4NaCl \tag{6.16}$$

$$UCl_4 + 4NaOH = U(OH)_4 + 4NaCl \tag{6.17}$$

The precipitates (Th–U concentrate contain 40–50% ThO_2, 11% U_3O_8, 8–10% RE_2O_3) are filtered, washed, and treated with concentrate HNO_3 to redissolve the uranium and thorium to form nitrate solution.

$$Th(OH)_4 + 4HNO_3 = Th(NO_3)_4 + 4H_2O \tag{6.18}$$

$$U(OH)_4 + 4HNO_3 = U(NO_3)_4 + 4H_2O \tag{6.19}$$

Solvent extraction with TBP (tributyl phosphate) is used to separate uranium and thorium from nitrate solution. Extraction with 10% TBP gets uranium in the organic phase leaving thorium and rare earths metals in the aquous phase. The raffinate is next contacted with 50% TBP to extract thorium in organic phase.

Scrubbing of impurities from the extract is done with dilute (1 N) HNO_3 and the thorium is stripped from the scrubbed extract by slightly acidified water [(0.2 N) HNO_3]. Thorium nitrate solution obtains in pure condition.

6.3.2 *Thorium Oxalate Formation*

Oxalic acid is reacted with thorium nitrate to precipitate of thorium oxalate (which contain Al-3, Ca-40, U-2 to 2.5, Ni-1, Cu-1, Fe-20, RE-40, Si-30, B-0.1, Cd-0.4, Zr-100 ppm).

$$Th(NO_3)_4 + 2H_2C_2O_4 = Th(C_2O_4)_2 + 4HNO_3 \tag{6.20}$$

6.3.3 *Chlorination of Thorium Oxalate*

The precipitate is heated for 2 h at 260 °C in argon atmosphere to remove moisture and to get dried thorium oxalate. The temperature is raised slowly to 350 °C and chlorine gas is than admitted into the reactor for converting thorium oxalate to thorium-oxy-chloride ($ThOCl_2$).

$$2Th(C_2O_4)_2 + 4Cl_2 = 2ThOCl_2 + 7CO_2 + CCl_4 \tag{6.21}$$

Chlorination is finally completed to $ThCl_4$ formation at 620 °C by passing a mixture of CCl_4–Cl_2 or CO–Cl_2 or CCl_4–CO_2.

$$2ThOCl_2 + CCl_4 + Cl_2 = 2ThCl_4 + CO_2 + Cl_2 (99\% \text{ yield}) \quad (6.22)$$

$$ThOCl_2 + CO + Cl_2 = ThCl_4 + CO_2 \quad (98\% \text{ yield}) \quad (6.23)$$

$$2ThOCl_2 + CCl_4 + CO_2 = 2ThCl_4 + 2CO_2 \quad (93\% \text{ yield}) \quad (6.24)$$

6.3.4 Purification of $ThCl_4$

$ThCl_4$ is prepared from oxalate usually containing some residual $ThOCl_2$ and unreacted ThO_2. For producing high purity metal, $ThCl_4$ should not contain oxygen-bearing compounds and carbon. Vacuum sublimation of $ThCl_4$ is done at 750 °C at argon pressure (25 mm of Hg). Distilled $ThCl_4$ is condensed on an air-cooled tube. The vacuum sublimed $ThCl_4$ is white, dense and crystalline.

6.3.5 Reduction of $ThCl_4$

(a) By Magnesium: $ThCl_4$ is reduced with Mg to get thorium sponge. Sufficient quantity of Mg (freshly pickled and acetone dried) is used to reduce $ThCl_4$ to form Th-20 wt% Mg liquid alloy. During the reduction, the temperature is maintained at 960 °C and a positive pressure of argon, kept arround 10 cm of Hg. Two hours heating is required for completion of the reaction. As the Th–Mg alloy is not completely liquid at this temperature, the temperature of the system is raised to 1050 °C to ensure good slag and alloy separation.

$$ThCl_4 + 3Mg = Th - Mg + 2MgCl_2 \quad (6.25)$$

Vacuum distillation is done to separate the thorium metal from Th–Mg alloy and $MgCl_2$. The charge is slowly heated under vacuum and the distillation is carried out at a holding temperature of 900 °C and final vacuum of 0.06 μ (6×10^{-5} mm of Hg) and holding time is 10 h (94% yield). After the vacuum distillation, the system is cooled to about 75 °C and then filled with argon for further cooling to room temperature. Before dismantling the system, the sponge (C-180, O-880, N-20 in ppm) is given a conditioning treatment with a mixture of air and argon to avoid sudden oxidation during handling.

b) By Sodium: The reaction between the $ThCl_4$ and Na is initiated by heating the mixture charge to about 525 °C, and the treatment is continued with subsequent heating at 800 °C for about 10 h.

$$ThCl_4 + 5Na = Th - Na + 4NaCl \tag{6.26}$$

The by-product, NaCl, and the excess Na from the charge are removed from the Th–Na alloy by heating in a vacuum at 900 °C for 16 h.

6.3.6 *Purification of Thorium Metal*

Thorium sponge is purified by electron beam melting at constant power level for different lengths of time to obtain thorium ingot (C-50, O-80, N-35 in ppm). The thorium sponge is then subsequently arc melted to get solid thorium ingot.

6.4 Production of Thorium Powder

(a) Reduction of ThO_2 by Ca: The reduction of ThO_2 by Ca in the presence of $CaCl_2$, according to the reaction:

$$ThO_2 + 2Ca + (2.4\,CaCl_2) = Th + 2CaO + (2.4\,CaCl_2) \tag{6.27}$$

The actual charge includes 3 mol excess of Ca. The mixed charge was heated to 950 °C in a closed steel bomb, under an inert gas atmosphere for 1 h during which the calcium reduces the thoria. The products are cooled and are leached with dilute acetic acid to dissolve the CaO and the unreacted Ca. The thorium metal remains as a powder and is washed successively with water, alcohol and ether. After that the powder is vacuum dried, it is compacted at 3.15–15.75 kg/mm^2 and then vacuum sintered at temperature up to 1450 °C for about 30 min to produce a dense ductile metal with purity better than 99.77%. The principal impurity is oxygen which is present about 0.15–0.17%.

Further improvement on this process is achieved by elimination of $CaCl_2$ in the reaction and use of a thin walled cup in place of the closed bomb, and reaction takes place under a static pressure of argon gas. Thorium powder (99.8%) is recovered by leaching the charge with dilute acetic acid, water, alcohol and ether; followed by vacuum drying.

(b) Fused Salt Electrolysis: Small scale production of pure thorium powder has been achieved by electrolysis of $ThCl_4$ in either NaCl or a mixture of NaCl and KCl. A high yield (90%) of good quality metal was produced by fused NaCl salt method, but preparation of the chloride from the nitrate feed material was found to be faster in the NaCl–KCl fused salt system. A bath with a 2:1 ratio of NaCl to KCl is heated at a temperature of 780–850 °C in a graphite crucible which served as anode. The cathode is made of Mo metal, which is suspended centrally in the bath; and electrolyte is the mixture of $Th(NO_3)_4$ and NaCl–KCl (chloride must be low in oxide and moisture content to obtain a pure product; also free from Fe, Ni, Zr, Cu, and Mn impurities, which are more electro-positive than thorium. These impurities are almost completely deposited on the cathode along with thorium).

$$Th(NO_3)_4 + 4NaCl/4KCl = ThCl_4 + 4NaNO_3/4KNO_3 \tag{6.28}$$

$$ThCl_4 = Th^{4+} + 4Cl^- \tag{6.29}$$

$$\text{At anode}: 4Cl^- - 4e = 2Cl_2 \tag{6.30}$$

$$\text{At cathode}: Th^{4+} + 4e = Th \tag{6.31}$$

Thorium powder is deposited on Mo cathode, the cathode along with deposit are slowly taken out from the electrolyte bath; deposits are washed with dilute H_2SO_4, and dried.

6.5 Production of Massive Thorium Metal

Thorium Fluoride Reduction: Thorium oxide is converted to the ThF_4 with anhydrous hydrofluoric acid (HF).

$$ThO_2 + 4HF = ThF_4 + 2H_2O \tag{6.32}$$

ThF_4 is mixed with a 20% excess of redistilled and sized (to remove the fines) Ca metal and with anhydrous $ZnCl_2$. The mixture is then charged to a refractory lined steel bomb. At about 640 °C the $ZnCl_2$ reacts exothermically with Ca and provides additional heat to the primary reaction.

$$ZnCl_2 + Ca = Zn + CaCl_2 \tag{6.33}$$

$$ThF_4 + 2Ca = Th + 2CaCl_2 \tag{6.34}$$

$$ThF_4 + 3Zn = Th - Zn + 2ZnF_2 \quad (6.35)$$

The reaction that ensues generates sufficient heat to fuse the products, Th–Zn alloy and slag. Zinc in turn reacts with the thorium at about 1200 °C to form alloy of low melting point; and the slag contains CaF_2–$CaCl_2$–ZnF_2. The dense molten alloy collects at the bottom of the bomb and under the thick layer of fused slag. Th–Zn alloy is recovered after the products have solidified in position and cooled to room temperature. This alloy is one large massive piece of metal, referred to as a *biscuit*, which is free of adhering slag. The biscuit is then placed in a graphite crucible and slowly heated for several hours under vacuum to a temperature of 900–1200 °C to distill the Zn. Holding the biscuit for 1 h at this temperature leaves a fraction of a percent of Zn in the sponge. Holding at this temperature for up to 8 h removes essentially all of the Zn.

Remelting of the sponge is done by induction furnace at 1800 °C; the main impurities in the casting are about 0.05% C (from the graphite moulds used in dezincing and in casting), about 0.02% Be (from bio-crucible), 0.02% N, about 1.25% oxide (as ThO_2) and trace amounts of Al, Fe and Si. Consumable electrode arc melting is also used to minimize contamination during the melting process. This gives lower O, Al, Si contents than in induction furnace process.

6.6 Properties

Pure thorium is a soft metal and paramagnetic in nature. The melting temperature is affected by small amounts of impurities, melting temperature range from a low of 1120 °C to a high of 1750 °C, boiling point is 4200 °C. When etched, its surface has a silvery-white lustre but darkens when exposed to air for prolonged periods. Thorium has a FCC structure at room temperature. At high temperature (about 1400 °C) allotropic transformation to the BCC structure takes place.

Crystal structure: (a) FCC → Room temperature to 1400 °C → a_0 = 5.086 Å at 25 °C

(b) BCC → 1400 °C to 1750 °C → a_0 = 4.11 Å at 1450 °C.

The atomic mass of thorium is 232.05. The theoretical density of the thorium at room temperature, based on the average lattice parameter of 5.086 Å, is 11,720 kg/m^3. Density of Ca reduced cast metal is between 11,500 kg/m^3 and 11,600 kg/m^3; and arc melted metal is 11,660 kg/m^3. Thermal conductivity as well as electrical resistivity increase with increasing temperature.

Amount as well as kind of impurities present have varying effects on the mechanical properties of thorium. Because of the chemical reactivity of hot thorium with oxygen and nitrogen of the air, it is necessary to have an irert gas protective atmosphere during hot fabrication. An important consideration in selecting a solid fuel material for a nuclear reactor regarding its corrosion behaviour with respect to the coolant. Ideally, the fuel should be fully corrosion resistant since it may be exposed to the coolant through failure of the protective cladding. Corrosion of thorium in air is increased with increasing temperature, above 400 °C there is a rapid increase in the corrosion rate.

Thorium forms a protective adherent film in boiling water (95–200 °C). Weight change is 1×10^{-4} mg/cm²/h.

$$Th + 2H_2O = ThO_2 + 2H_2 \quad (6.36)$$

$$2Th + 3H_2 = 2ThH_3 \quad (6.37)$$

$$2ThH_3 + 4H_2O = 2ThO_2 + 7H_2 \quad (6.38)$$

Most thorium alloys have less corrosion resistance than unalloyed thorium in high temperature water. The Th–Zr alloys show an increased resistance in high temperature water. In water at 200 °C, the rate of weight loss is least for Th–10% Zr alloy. Alloys with small amounts of C, Si, Ti and Zr have better corrosion resistance than pure thorium in high temperature water.

Ingots of commercial grade thorium can be readily worked either hot or cold by most mechanical fabrication processes. The FCC metal has low hardness, low strength, and high ductility at room temperature. During cold rolling, drawing of thorium without intermediate annealing, the metal increases in hardness only moderately up to 40% reduction in area. Futher cold rolling up to 99% reduction shows very little increase in hardness. The chemical reactivity of thorium must be considered when heating for hot working. The heating may be carried out in a fused salt bath, or the thorium metal may be jacketed with another metal such as copper. Thorium metal can be machined easily.

6.7 Applications

(a) Nuclear use: Thorium is important as a potential source of atomic energy. Just as natural uranium-238 can, by absorption of a neutron, be converted into a new element plutonium, of critical value in the release of atomic energy, so thorium-232 can become a new isotope of uranium-233. An important difference exists, however, in that natural uranium-238 contains its own nuclear fuel; thorium having no naturally occurring isotopes has no such advantage and therefore in any thorium atomic energy pile, fuel in the shape of plutonium or uranium-235 has to be supplied. For this reason thorium cannot yet compete with uranium in the production of atomic energy. The major applications of thorium are in nuclear power. The form of the thorium for use in nuclear applications includes the metal, alloys and compounds. Thorium is used to form U^{233}.

(b) Non-nuclear use: Thorium is an important alloying element in magnesium because it contributes high strength properties and creep resistance to magnesium at elevated temperature. Although thorium metal has a density similar to lead, the amounts necessary to impart these desirable properties to magnesium are sufficiently low (about 3%) to give light-weight alloys that are definitely advantageous for use in aeroplane and missile construction. Thorium metal is used as a non-energy application, like use as a low density alloy (a few percent Th in Mg) with superior high temperature properties.

Other non-nuclear uses of thorium include application in several lamps of the gaseous discharge type, due to its high electron-emission properties. The metal has been used in photoelectric tubes for the measurement of a wide band of the ultraviolet spectrum, and it is also used in germicidal lamps of the cold cathode type.

Part II
Extractive Metallurgy of Common Metals

Sources of Metals

There are mainly two types of source of metals:

1. **Primary sources of metals** → These are further classified as follows:
   (a) Earth's crust → most of the metals in the form of ore or minerals
   (b) Sea water → Mg, Na, K etc are in solution
   (c) Sea nodules → most of the metals (Mn, Ni, Fe, Cu, Co etc) are found in sea nodules, they are lumps of material, formed of concentric layers of metallic hydroxides around a core.
2. **Secondary sources of metals** → scrap from different metals industries or from different by-products.

Non-ferrous Mineral Wealth of the World

Table II.1 shows the minerals of common non-ferrous metals in the world. This non-ferrous mineral wealth can be classified as follows:

1. Large quantity: minerals containing Al, Ti, Mn, Mg, Cr, Be, Zr, Th and rare earths metals. If properly exploited they can meet domestic demand and also be exported.
2. Medium quantity: minerals containing Cu, Pb, Zn, Ni, Sn, Cd, Au, V, U. Insufficient to meet growing domestic demand and hence should be fully exploited.
3. Very small quantity: minerals containing Sb, Bi, B, Co, Hg, Mo, Nb, Ta, Ag, W. Insufficient for commercial exploitation.

Table II.1 Minerals of common non-ferrous metals in the world

Metal	Mineral	Chemical formula	Main location
Aluminum (Al)	Diaspore Gibbsite Kaolinite	$Al_2O_3{\cdot}H_2O$ $Al_2O_3{\cdot}3H_2O$ $Al_2O_3{\cdot}2SiO_2{\cdot}2H_2O$	Australia, Guinea, Brazil, Jamaica, and India
Antimony (Sb)	Stibnite	Sb_2S_3	China, Bolivia, Russia, Mexico
Beryllium (Be)	Beryl	$3BeO{\cdot}Al_2O_3{\cdot}6SiO_2$	Zimbabwe, USA, and India
Bismuth (Bi)	Bismite Bismuthinite Bismutite	Bi_2O_3 Bi_2S_3 $Bi_2O_3{\cdot}CO_2{\cdot}H_2O$	Peru, USA, Mexico, Japan, South Korea, and Canada
Boron (B)	Colemanite Tincal Kernite Ulexite	$2CaO{\cdot}3B_2O_3{\cdot}5H_2O$ $Na_2O{\cdot}2B_2O_3{\cdot}10H_2O$ $Na_2O{\cdot}2B_2O_3{\cdot}4H_2O$ $Na_2O{\cdot}2CaO{\cdot}5B_2O_3.16H_2O$	USA, Italy, and South America
Cadmium (Cd)	Native Cadmium Greenockite	Cd CdS	Siberia
Chromium (Cr)	Chromite	$FeCr_2O_4$	South Africa, and Zimbabwe
Cobalt (Co)	Cobaltite	CoAsS	Zaire, Zambia, New Caledonia, Australia
Copper (Cu)	Native copper Chalcocite Covellite Chalcopyrite Cuprite	Cu Cu_2S CuS $CuFeS_2$ Cu_2O	USA, Chile, Canada, Russia, Peru, and Zambia
Gold (Au)	Native gold Calaverite Sylvanite	Au $AuTe_2$ $(AuAg)Te_2$	USSR, Australia, South Africa, Ghana, Zaire, Zimbabwe, Canada, and USA
Lead (Pb)	Galena Cerussite Anglesite	PbS $PbCO_3$ $PbSO_4$	Russia, Australia, Canada, Peru, Yugoslavia, South Africa, Sweden, Spain, and Japan
Magnesium (Mg)	Magnesite Dolomite	MgO $MgCO_3{\cdot}CaCO_3$	All over world
Manganese (Mn)	Pyrolusite Rhodochrosite Rhodonite	MnO_2 $MnCO_3$ $MnSiO_3$	South Africa, Russia, Brazil, Ghana, and India
Molybdenum (Mo)	Molybdenite Molybdite	MoS_2 MoO_3	USA, Russia, Canada, and Chile
Nickel (Ni)	Garnierite Nicolite Pentlandite	$(Ni, Mg)SiO_3{\cdot}xH_2O$ NiAs $(Fe, Ni)_9S_8$	Canada, Russia, New Caledonia
Niobium (Nb) and Tantalum (Ta)	Columbite Pyrochlore	$(Fe, Mn)(Nb, Ta)_2O_6$ $(Na,Ca)_2(Nb,Ti)(OF)_7$	Zaire, Brazil, Nigeria, Malagasy, Norway, Sweden, Russia

(continued)

Table II.1 (continued)

Metal	Mineral	Chemical formula	Main location
Silicon (Si)	Quartz Quartzite Chalcedony	SiO_2 SiO_2 SiO_2	All over world
Silver (Ag)	Native silver Argentite Cerargyrite	Ag Ag_2S AgCl	Mexico, Bolivia, USA, and Canada
Tin (Sn)	Cassiterite Stannite	SnO_2 $(Cu_2S{\cdot}FeS{\cdot}SnS_2)$	China, Thailand, Malaysia, and Bolivia
Titanium (Ti)	Ilmenite Rutile	$FeO{\cdot}TiO_2$ TiO_2	Brazil, Canada, India, Australia, and USA
Tungsten (W)	Wolframite Scheelite	$FeWO_4$ $CaWO_4$	China, Canada, Russia, North Korea, South Korea, USA
Vanadium (V)	Patronite Carnotite Vanadinite	V_2S_5 + Sulphur $K_2O{\cdot}2UO_3{\cdot}V_2O_5{\cdot}3H_2O$ $3Pb_3(VO_4)_2PbCl_2$	Russia, and South Africa
Zinc (Zn)	Sphalerite Zincite Franklinite	ZnS (Zinc blende) ZnO (Fe, Zn, Mn)O	Australia, Burma, Yugoslavia, Spain, and Italy

Chapter 7
Copper

7.1 Introduction

The end of the eighteenth century saw Swansea as the largest copper (Cu) producer in the world. The technique of modern copper metallurgy owes a great deal to the early work done in Wales. The reverberatory, which still plays a very important part, was largely developed there; and later, Welsh smelters brought their experience to bear in all parts of the world. The nineteenth century saw the decline of Welsh copper smelting. America gradually assumed ascendancy in the art of copper smelting and continued in the lead.

Copper is the most important non-ferrous metal. Not only does it form extremely useful alloys with metals such as Zn and Sn, but it also has unique properties (such as being a good conductor of electricity) in the unalloyed state.

7.2 Sources

Copper exits in nature mostly in the form of copper sulphide with or without other sulphides. The sulphide minerals are the predominant feature in copper metallurgy, yielding over 90% of the copper supply of the world. Deposits are in general low grade, the majority being worked today containing 2% copper or less. The common minerals of copper (figures within the parentheses give the chemical composition and percentage of theoretical copper contained): (a) chalcopyrite ($CuFeS_2$, 34.5% Cu); (b) chalcocite (Cu_2S, 79.8% Cu); (c) bornite (Cu_5FeS_4, 55.5% Cu); (d) covellite (CuS, 66.5% Cu); (e) enargite ($CuAsS_4$, 48.4% Cu); (f) malachite ($CuCO_3.Cu(OH)_2$, 57.3% Cu); (g) azurite ($2CuCO_3. Cu(OH)_2$, 55.1% Cu); (h) cuprite (Cu_2O, 88.8% Cu); and (i) chrysocolla ($CuSiO_3.2H_2O$, 36.2% Cu). Chalcopyrite is the most commercial ore. Minerals theoretically contain large amount of copper, but actually ores contain only small amounts of copper due to presence of other

S.K. Dutta and D.R. Lodhari, *Extraction of Nuclear and Non-ferrous Metals*,
Topics in Mining, Metallurgy and Materials Engineering,
https://doi.org/10.1007/978-981-10-5172-2_7

sulphides and gangue minerals. Actually, a sulphide ore that contains 0.5–2% Cu is considered as good for extraction after concentration.

7.3 Extraction of Copper

The extraction of copper by the convention route (concentration, roasting, smelting, converting and refining) is shown in Fig. 7.1. The main steps involved are as follows:

1. Enrich the ore by concentration.
2. Roast for partial elimination of sulphur contained in the copper concentrate.
3. Smelt in a reverberatory furnace to concentrate the copper into matte and to eliminate the gangue materials as slag.
4. Convert matte to blister copper, which is cast as anode.
5. Electrolytically refinine impure copper.

7.3.1 Concentration

A sulphide ore normally contains 0.5–2% Cu. The ore is crushed and ground up to 40 μm for froth flotation. The p_H of froth flotation is controlled by the addition of CaO; xanthate reagent is used as a collector. A differential flotation is used to selectively recover copper sulphide from the mixture of copper sulphide, lead sulphide and zinc sulphide. The concentration of copper sulphide contains 15–35% Cu, 15–35% Fe, 25–35% S and 3–35% gangue.

7.3.2 Roasting

The object of roasting is to burn off part of the sulphur and the purpose of roasting is to partially oxidize the iron sulphide present in the copper sulphide concentrate in order to remove it in the form of a slag at the smelting stage. The affinity between copper and sulphur is greater than between sulphur and any other metals present. In smelting, the copper will take enough of the sulphur to form Cu_2S, the residual sulphur combining with the iron to form FeS. The amount of FeS, which will accompany the Cu_2S in the matte, is entirely dependent on how much sulphur is left to combine with the iron after all the copper has been satisfied; the amount being controllable by the preliminary roasting, which in turn controls the grade of matte.

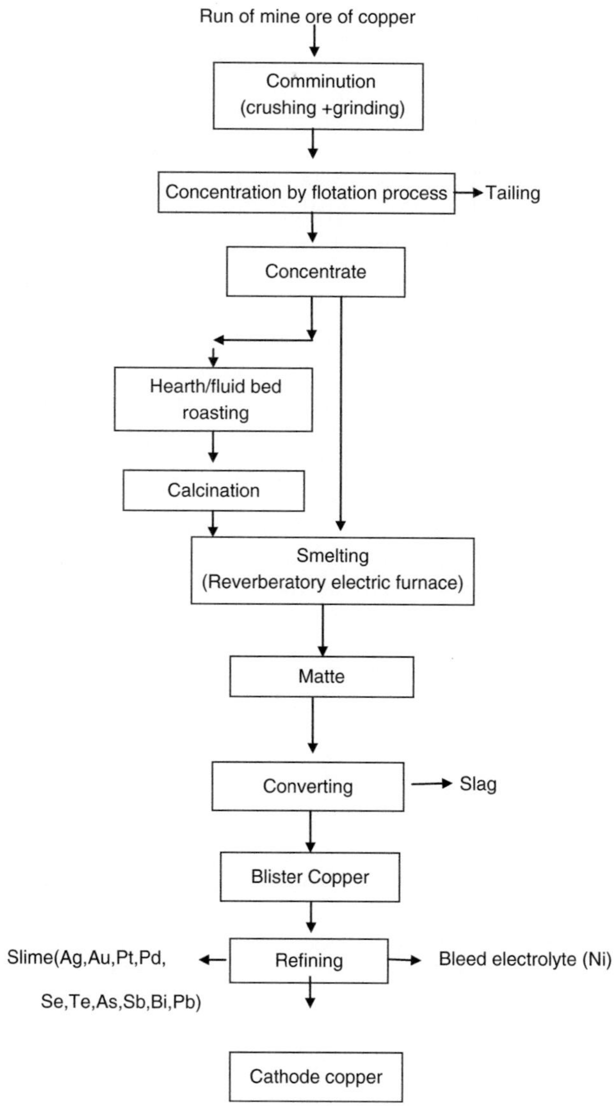

Fig. 7.1 Flow diagram for copper extraction

The extent of roasting determines the copper grade of the matte produced in the smelting stage. If a high grade matte is required, the roasting is carried to a high degree, and for a low grade matte the roast would not be pushed to the same extent. The more usual grade of matte ranges from 35–45% copper. The highest grade of matte (i.e. the one containing the greatest concentration of copper) is not necessarily the most economic to aim at as it requires a high degree of roasting and gives rise to high costs and dust. Also, copper losses in smelting tend to follow the grade of matte; in other words the higher the grade of matte, the higher the copper losses in the slag.

For concentrates with a high iron sulphide content and a low copper sulphide content (< 25% Cu) roasting is necessary; but for high grade copper concentrates (> 30% Cu), roasting is not necessary. The bulk of sulphur elimination is due to the oxidation of ferrous sulphide and not to the sulphide of copper, any excess sulphur after the formation of the Cu_2S is expelled. Complete elimination of sulphur is neither sought nor desired, the oxidation of the iron supply provides sufficient heat in the roasting operation. Sulphates will form at low temperature with excess of air and sulphur dioxide, but at higher temperature will react with sulphides, resulting in the formation of oxides. High temperature and low sulphur dioxide concentration favour the formation of oxides (like Fe_2O_3, Fe_3O_4 etc.).

Roasting is carried out in multiple hearth roasters, the temperature of each hearth is gradually increased from room temperature to 550°C. During roasting, several reactions take place simultaneously as follows:

$$2CuFeS_2 + 6.5O_2 = 2CuO + Fe_2O_3 + 4SO_2 \tag{7.1}$$

$$CuFeS_2 + 4O_2 = CuSO_4 + FeSO_4 \tag{7.2}$$

$$2CuFeS_2 + O_2 = Cu_2S + 2FeS + SO_2 \tag{7.3}$$

$$2CuFeS_2 + 4.5O_2 = Cu_2S + Fe_2O_3 + 3SO_2 \tag{7.4}$$

$$2CuFeS_2 + 7.5O_2 = CuO.CuSO_4 + 2FeSO_4 + SO_2 \tag{7.5}$$

$$3CuFeS_2 + 9.5O_2 = 3CuO + Fe_3O_4 + 6SO_2 \tag{7.6}$$

$$6CuFeS_2 + 13O_2 = 3Cu_2S + 2Fe_3O_4 + 9SO_2 \tag{7.7}$$

The roasted calcine (which consists of sulphides of copper and iron, oxides of iron, and mixed sulphides of copper and iron) is transferred to the smelting furnace in the hot condition.

7.3.3 Smelting

The sulphur elimination is bound up with smelting and in fact determines the grade of matte. To produce a high grade matte in the reverberatory furnace (as shown in Fig. 7.2) necessitates a high volume of slag which in turn is likely to lead to higher copper losses and possibly losses of precious metals which normally accompany the product of smelting, matte. The matte is not a true chemical compound, but rather a varying mixture of the two sulphides Cu_2S and FeS.

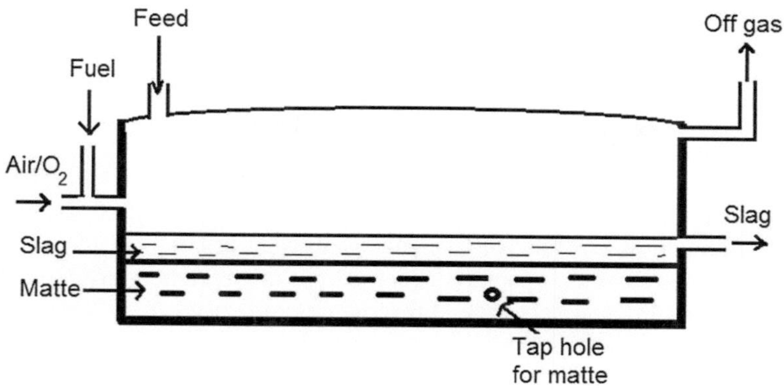

Fig. 7.2 Reverberatory furnace for copper smelting

Smelting separates the metal sulphides in roasted calcine or concentrate from the gangue. Such separation can be achieved by smelting the charge with suitable flux at a temperature of 1250°C. Two layers of liquid are formed, an upper slag layer (specific gravity in the range of 2.8–3.8) consists of gangue and flux; lower layer contains matte (specific gravity in the range of from 5.0–5.5) which consists of sulphides of copper and iron (35–45% Cu, 20–22% S and 25–35% Fe). Due to difference in specific gravities, a clear-cut separation of layers is observed.

During smelting, exchange reactions take place between the oxide/sulphate of copper and the iron sulphide, due to oxygen having a higher affinity for iron than copper:

$$6CuO + 4FeS = 3Cu_2S + 4FeO + SO_2 \tag{7.8}$$

$$2CuSO_4 + 2FeS = Cu_2S + 2FeO + 3SO_2 \tag{7.9}$$

$$Cu_2O + FeS = Cu_2S + FeO \tag{7.10}$$

The unoxidized iron sulphide reduces the higher iron oxides to lower oxides:

$$10Fe_2O_3 + FeS = 7Fe_3O_4 + SO_2 \tag{7.11}$$

$$3Fe_3O_4 + FeS = 10FeO + SO_2 \tag{7.12}$$

The resultant ferrous oxide (FeO) reaction with gangue (SiO_2) present in the concentrate/calcine forms iron silicate slag during smelting:

$$FeO + SiO_2 = FeO.SiO_2 \tag{7.13}$$

In the presence of lime (as flux), iron-calcium silicate slag is produced:

$$FeO.SiO_2 + CaO = FeO.CaO.SiO_2 \qquad (7.14)$$

The melting point of the slag formed is about 1150°C, and to obtain a fluid slag with a low copper content, smelting is carried out at a temperature of 1250°C. The net result of the smelting operation is matte (containing Cu_2S and FeS), slag consisting largely of iron silicate, and sulphur dioxide gas.

Generally, the smelting operation is carried out in a reverberatory furnace fired with coal or oil. Thermal efficiency of reverberatory furnace is very low (30%) due to outgoing gases carrying a lot of heat. To improve heat utilization, outgoing gases are passed through air heaters, the efficiency then rising to 55–60%.

Submerged electric arc furnace is also used for smelting. Electric furnace is more advantageous, where the cost of power is low, than reverberatory furnace; because the generation of a large volume of combustion gases is avoided. Advantages of electrical smelting are low energy consumption, high thermal efficiency, steady operation, long life of refractory (basic) roof and lining, and a small volume of gases from which sulphur dioxide may be economically recovered.

7.3.4 Converting

The main purpose of converting is to remove iron, sulphur and other impurities from matte. The molten matte, produced by smelting operation, is charged into a side blown converter (as shown in Fig. 7.3). In the converter, which is lined with the chrome-magnesite refractory, the atmosphere is highly oxidizing compared with the neutral or mildly oxidizing atmosphere during smelting operation. Air or oxygen-enriched air (up to 32% by volume) is injected into the molten matte through side tuyeres. The isolation of the copper from the matte is achieved by a progressive oxidation in a converter. When air is blown through molten matte, iron sulphide is oxidized before the copper sulphide.

Slagging stage: In the slagging stage, iron sulphide present in the matte is oxidized and the oxide is slagged out by the addition of siliceous flux. Part of the iron oxide (FeO) is also oxidized to higher oxide:

$$2FeS + 3O_2 = 2FeO + 2SO_2, \quad \Delta H^\circ = -460,826\,J/mol\,of\,FeS \qquad (7.15)$$

$$2FeO + SiO_2 = 2FeO.SiO_2(Fayalite) \qquad (7.16)$$

$$6FeO + O_2 = 2Fe_3O_4 \qquad (7.17)$$

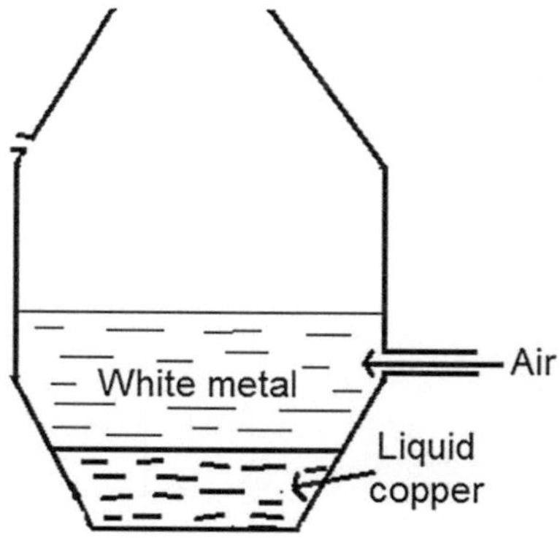

Fig. 7.3 Side blown converter for copper production

In the converter the heat, necessary to keep the reaction going on has to be provided by the oxidation of the iron sulphide [reaction (7.15)]. A high grade matte (containing low iron) therefore has difficulty in maintaining the heat essential for the conversion. Primary oxidation starts and slag begins to be produced, oxidation furnishing the heat necessary for the slagging action.

The slag consists of 40–50% Fe, 20–30% SiO_2, 1–5% (CaO + MgO) and 2–9% Cu. The copper is recovered from the slag by transferring the slag to the smelting furnace. The iron oxide (FeO) is present in the slag mostly as fayalite saturated with magnetite (Fe_3O_4). The magnetite forms a coating on the chrome-magnesite refractory, which prevents slag attack on the refractory. Slagging is carried out in stages: (a) first freshly obtained matte is added to the converter and then air blowing is started; (b) slag formed is skimmed off by tilting the converter. The molten slag and matte phases are immiscible.

During converting, the oxidation of the iron sulphide produces sufficient heat [reaction (7.15)] to overcome the heat lost to the surroundings and to maintain the matte and slag in molten state. For a large converter where the blowing rate as well as the iron sulphide oxidation rate are high, heat generation is hence very high as well; this extra heat is utilized for melting of copper scrap which is generated in the plant.

After an hour or so the reaction is completed, air turned off, the converter tilted and the slag poured off into a ladle. More matte and flux are added, air readmitted and the action recommences. The cycle of operations, charging and skimming of slag proceeds until sufficient copper sulphide (white metal) has been accumulated in the converter to complete the process by converting the white metal to blister copper [reaction (7.20)]. The time taken to run down matte and produce blister copper depends on a number of factors: (a) grade of matte and flux; (b) condition of lining, etc.

Blister formation stage: After slagging is completed, the converter essentially contains Cu_2S which is called "white metal" due to its appearance. Cu_2S is oxidized to form copper by the following two reactions:

(i) Copper oxide is formed:

$$2Cu_2S + 3O_2 = 2Cu_2O + 2SO_2 \quad (7.18)$$

(ii) Copper oxide reacts with further remaining copper sulphide to produce blister copper:

$$Cu_2S + 2Cu_2O = 6Cu + SO_2 \quad (7.19)$$

The overall reaction (combination of Eqs. 7.18 and 7.19):

$$3Cu_2S + 3O_2 = 6Cu + 3SO_2, \quad \Delta H^\circ = -217,317\,J/mol\,of\,Cu_2S \quad (7.20)$$

Reaction (7.20) shows the cuprous sulphide is reduced to blister copper without using any reducing agent; i.e. converting the white metal to blister copper. The final stages have to be given good attention owing to the chance of oxidation of copper. Boiling of the bath takes place due to evolution of sulphur dioxide gas from the molten bath. The heat necessary to maintain the charge is supplied by the oxidation of the iron and copper sulphides (reactions 7.15 and 7.20), the iron sulphide contributing by far the greater part. The amount of iron sulphide present is thus an important factor in the control of the operation. The highlights of the operation are as follows:

1. Production of white metal (Cu_2S) by oxidation of the FeS in the matte and its elimination as slag by combination with silica; and
2. Production of blister copper by further oxidation.

Air is blowing to the converter at a pressure of 9.842×10^{-3} kg/mm^2. The amount of air required for the conversion of matte to blister copper is found to be in excess of that calculated and in fact amounts to some 150–160%. This is due to the following reasons:

1. Leakage through the lining.
2. Formation of oxides other than the calculated FeO.
3. Air escaping uncombined with the gases through the mouth of the converter.

Two important factors for slag formation are (a) temperature and (b) flux. The formation of slag requires a minimum temperature in the region of 1250°C. Lower temperature results in the formation of Fe_3O_4 which carries copper into the slag; on the other hand, too high a temperature shortens the life of refractory (magnesite) lining. The function of flux (i.e. silica) is to combine with FeO, which is formed by oxidation of iron in the matte, to form slag. The composition of the slag also depends on the amount and type of flux. The converter slag contains 5% Cu, which is returned to the reverbatory furnace where it mixes with the molten charge and is cleaned.

From the Cu–Cu_2S phase diagram (Fig. 7.4), we understand that when white metal (Cu_2S) is oxidized and the amount of sulphur in the bath decreases to below 19.8%, the bath splits into two layers, i.e. a top sulphide layer (less dense) and a bottom copper layer (more dense) containing 1.2% sulphur. As the oxidation continues the volume of the sulphide layer decreases and the volume of the copper layer increases. When the sulphur level reaches 1.2%, only the metallic copper phase remains in the converter. At this stage, care must be taken to avoid over-oxidation of metallic copper. The blistery (bubble-like swelling, i.e. small holes) appearance on the surface of the sample of copper gives the name blister copper to the product, which contains 0.02–0.05% S along with 0.2–0.5% O. During solidification, these dissolved S and O in metal react to form SO_2 gas ($\underline{S} + \underline{O} = SO_2$). Due to evolution of gas from the metal the surface become blistery.

7.3.5 *Refining*

The blister copper is unsuitable for direct commercial application owing to the amount of impurities (consisting of arsenic, oxygen, sulphur, bismuth, iron, gold, silver etc., and totalling usually from 0.5% to 1.5%). Many impurities, including gases, affect adversely the mechanical properties of copper, in particular, its

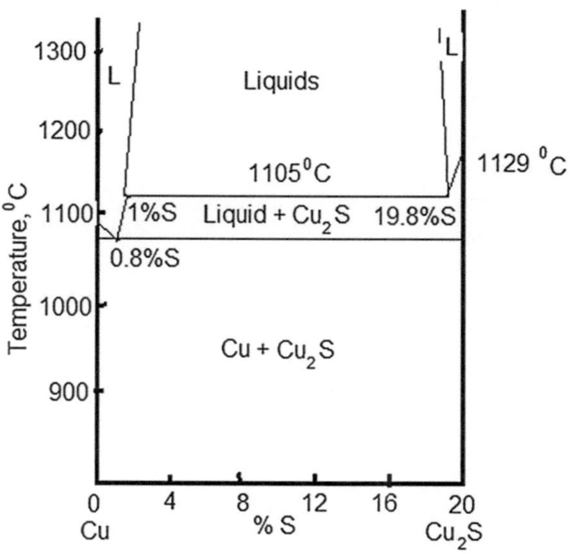

Fig. 7.4 Cu–Cu_2S phase diagram

plasticity. All impurities, except for the noble metals, have a detrimental effect upon electric conductivity of copper. Refining is therefore essential before the copper can be obtained in a suitable form for marketing. There are two purpose of refining: (a) to obtain the pure metal (for a higher degree of electrical conductivity), and (b) to recover the valuable precious metals.

7.3.5.1 Fire Refining

Refining involves the following stages: melting, oxidation of impurities and removal of slag, deoxidation (poling) and casting. Impurities are oxidized by air which is injected into molten copper through refractory-coated iron pipes. Oxidation takes place at the surface of air bubbles that float in the molten metal. Copper and all impurities, except the noble metals, are oxidized.

$$4Cu + O_2 = 2Cu_2O \tag{7.21}$$

Cuprous oxide is soluble in molten copper and rapidly diffuses throughout the bath. Oxidation of impurities (particularly metals, M) may be described by the general equation:

$$M + Cu_2O = MO + 2Cu \tag{7.22}$$

As per the Ellingham diagram (Fig. III.1, p. 176), impurities (such as Al, Si, Mn, Zn, Sn, Fe, Ni, As, Sb, Pb) have a greater affinity for oxygen than copper; therefore they all are oxidized by cuprous oxide [as per reaction (7.22)]. Oxides of impurities are insoluble in molten copper; they float to the surface and form slag which is skimmed off. Sulphur, present in copper in the form of dissolved Cu_2S, is oxidized according to the reaction (7.19).

To remove sulphur from liquid blister copper as SO_2 gas by oxidation with air and to eliminate oxygen by introducing hydrocarbons, the surface of the blister copper is oxidized at frequent intervals. The furnace doors are kept open in order to allow a constant mild blast of air. As a result of this slow oxidation process (2–4 h), impurities such as S, Fe, Se and Zn are oxidized and the solid oxides rise to the top where they are skimmed off. Some copper is also oxidized during the process; the process is discontinued when the level of Cu_2O in the bulk of the metal reaches 6%. At this stage, almost all sulphur is eliminated from the metal. Slag should be thoroughly removed before copper is deoxidized; otherwise the impurities oxides will be reduced from the slag and again go to the molten copper. The bath is then skimmed clean of slag, covered with charcoal or coke, and poling is commenced. Once the oxidation is completed, Cu_2O is reduced by poling with green woods. On ignition, these woods generate hydrocarbons which reduce the Cu_2O. Vigorous stirring take place at the molten bath by bubbling of gases.

$$4Cu_2O + CH_4 = CO_2 + 2H_2O + 8Cu \tag{7.23}$$

Pitch virtually refers to the oxygen content which is usually within the range of 0.02–0.1%. The *set* or appearance of the surface of a button sample taken from the furnace is a rough guide to the oxygen content. If too much oxygen is present, the set is concave; if too little, the set is convex. When oxygen is correctly positioned, the set is level. Other factors such as sulphur dioxide are, however, liable to affect set. Poling is continued until sample shows a level set indicating that oxygen content of 0.03–0.06% has been attained, this being the normal oxygen content of tough pitch copper. The purity of fire-refined copper is generally around 99.7%.

7.3.5.2 Electrolytic Refining

Fire refining method does not effectively remove such impurities as bismuth, tellurium, selenium and nickel; nor does it recover the precious metals. Electrolysis was introduced to make good these deficiencies. Fire-refined copper can be further refined by electrolysis. An electrolytic cell is 3–5 m long, 1.0–1.1 m wide and 1.0–1.3 m deep; design of cell is based on the utilization of minimum space with maximum cathode and anode areas. Fire-refined copper is cast in the form of an anode that weighs about 250–320 kg. Suspended parallel to the anodes are connected to the positive poles of direct current. The cathode is made of a pure copper or stainless steel sheet; the electrolyte consists of copper sulphate (35×10^{-3} kg of Cu per litre), sulphuric acid (0.2 kg per litre), and some addition agents such as glue and alcohol. Electrolysis is carried out using current density of about 200 A/m^2 at a temperature of 50–60°C. When a direct current is passed through a solution of copper sulphate, anode copper goes into solution according to reaction (7.24), the liberated copper is deposited on the cathode. The anode plate will show a decrease in weight, due to copper going into solution, and the cathode a corresponding increase, due to copper having been deposited thereon.

During electrolysis, copper is transferred from the crude anode to the pure cathode as follows:

$$\text{At anode}: \mathrm{Cu} \rightarrow \mathrm{Cu}^{2+} + 2\mathrm{e} \tag{7.24}$$

$$\text{In the electrolyte}: \mathrm{CuSO_4} \rightarrow \mathrm{Cu}^{2+} + (\mathrm{SO_4})^{2-} \tag{7.25}$$

$$\text{At cathode}: \mathrm{Cu}^{2+} + 2\mathrm{e} \rightarrow \mathrm{Cu} \tag{7.26}$$

Impurities in the blister copper such as Fe, Co, Ni, Se, and Te go into the solution and the precious metals are collected just below the anode, which is known as anode slime. Impurities (like gold, and silver) are more positive than copper. They are settling out as anode slime. Anode slime consist of 10–40% Cu, 2–20% Pb, 0.5–5% As, 0-4% Te, 0–25% Ni, 2–10% S, 0.5–2% Fe, 0–25% Se and 8.5–14 kg of Ag/tonne, 0.25–1.7 kg of Au/tonne. The exact composition depends on the composition of the anode and the various, previous, metallurgical operations.

Cathode deposit is contaminated mainly by entrapped electrolyte and floating slime. To minimize this, part of the electrolyte is periodically removed from the baths and purified. Regeneration (i.e. purification) of electrolyte is necessary to avoid build-up of impurities and to remove excess copper. Copper accumulates in the electrolyte through partial chemical reaction involving participation of air or oxygen.

$$2Cu + O_2 + 2H_2SO_4 = 2CuSO_4 + 2H_2O \quad (7.27)$$

If impurities content is low and excess copper is only to be extracted, the electrolyte is processed by electrolysis in standard tank, but furnished with insoluble lead anodes. Copper of the electrolyte plates out at the cathodes in the process, oxygen evolves at the lead anodes, while some free sulphuric acid forms in the solution. Once copper is partially extracted, the electrolyte may be returned to the circulation system.

Actually, not all of the current is used to deposit the copper, and some current is lost because of short circuits between electrodes, current leakage and discharge of impurities' ions.

Some difficulties are encountered during electrolysis:

(a) The concentration of Cu^{2+} ions in the electrolyte decreases. As a result, water is electrolyzed and releases hydrogen which lowers the current efficiency.
(b) The concentration of Cu^{2+} ions in the electrolyte increases, and $CuSO_4$ may tend to be crystallized in the electrolyte.
(c) The copper sulphate content of the electrolyte increases beyond a certain limit, resulting in dissolution of Cu_2O (blister copper).

These difficulties can be overcome by the proper stirring, circulation of the electrolyte; and by using an inert anode made of lead. This procedure brings down the $CuSO_4$ content of the electrolyte. Pure copper (99.99%) is removed from the electrolyte after removal of impurities (such as As, Sb, Bi, and Ni). Arsenic, antimony, and bismuth are removed during the last stages of copper refining with the inert lead anode. Nickel is removed together with iron and cobalt as sulphate form.

7.4 Newer Processes

The flow diagram of the newer processes is shown in Fig. 7.5.

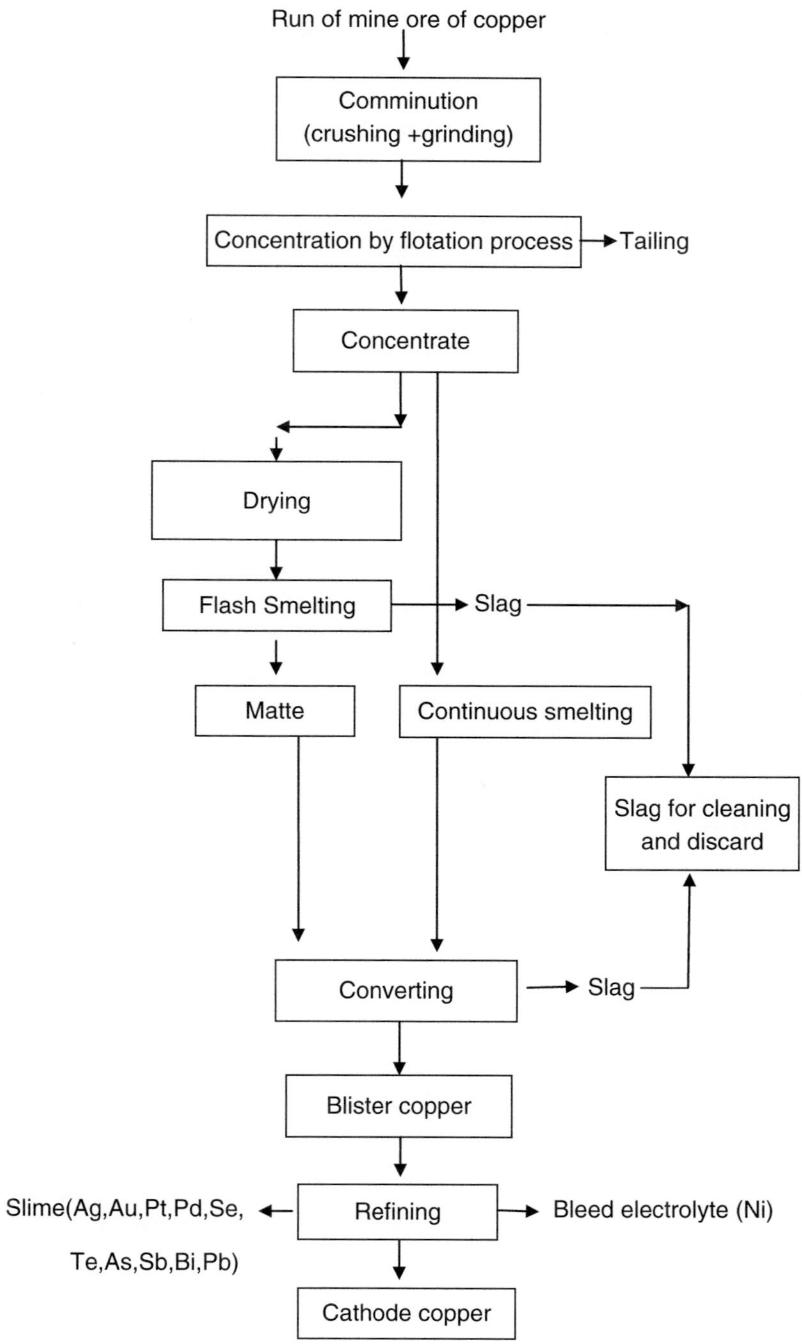

Fig. 7.5 Extraction of copper by other method

7.4.1 Flash Smelting Process

(a) The flash smelting process is the combination of two operations like roasting and smelting. Dried concentrate of copper and sand along with oxygen are injected into the flash smelter. The iron sulphides are combusted and generate the heat to smelt the concentrate to get matte, which is converted into the blister copper in converter. This process was developed by the International Nickel Company of Canada (INCO) and was operated commercially from 1953. Figure 7.6 shows the INCO flash smelting furnace. Most of the reactions are completed in the flame of the flash burner that leads to a high smelting rate (2.54 tonne/day/m^3). INCO had developed a new design for the flash burner which could achieve faster smelting rate (17.56 tonne/day/m^3). The new flash burner could complete smelting within two seconds and the composition of the smelter is almost equal to the value at equilibrium. No external fuel is used in this process; the matte contains about 45% Cu and slag contains only about 0.5–0.6% Cu. The off-gas of the INCO flash smelter contains about 80% SO_2 which can be used for the production of sulphuric acid.
(b) Another flash smelting process was developed around the same period by Outokumpu in Finland by using air, oxygen-enriched air or preheated air (Fig. 7.7). The matte produced in the Outokumpu flash smelting process contains 45–65% Cu depending on the quantity of the fuel (i.e. oil) and the degree of oxygen enrichment used. The smelting rates for preheated air and for enriched oxygen are 2.61 tonne/day/m^3 and 3.88 tonne/day/m^3 respectively. The slag contains high Cu content, i.e. 0.8…1.5% Cu.

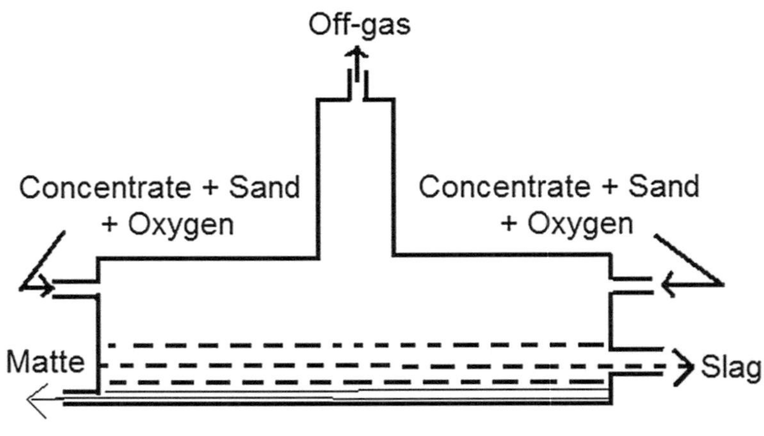

Fig. 7.6 INCO flash smelting furnace

Fig. 7.7 Outokumpu flash smelting furnace

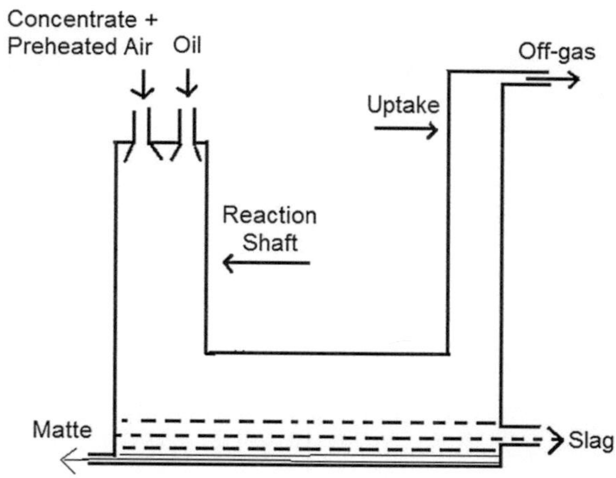

7.4.2 *Continuous Process*

Continuous process is the combination of flash smelting and converting operations in a single vessel; the copper concentrate can be charged at one end and copper metal can be continuously tapping from the other end. Continuous smelting processes developments—namely, the WORCRA process (developed in Australia) and the Noranda process (developed in Canada)—were patented in 1967. Mitsubishi (in Japan) has developed a process in which smelting and converting are carried out in separate containers; however, there is a continuous flow of matte to the converter from smelter and a continuous flow of blister copper from converter.

7.4.2.1 WORCRA Process

The name WORCRA is derived by combining the first three letters (i.e. WOR) of the inventor's name (H.K. Worner) and CRA (Conzinc Riotinto of Australia Limited). The WORCRA process is based on the conservation of energy obtained during smelting and converting of sulphide concentrates by integrating several efficient continuous unit operations. In the WORCRA process, a single furnace combines the following continuous operations which occur in separate but inter-connected zones:

1. Smelting.
2. Converting.
3. Slag cleaning by conditioning and settling.

Figure 7.8 shows the three operations in a horizontal plane. The WORCRA process is characterized by the following main features:

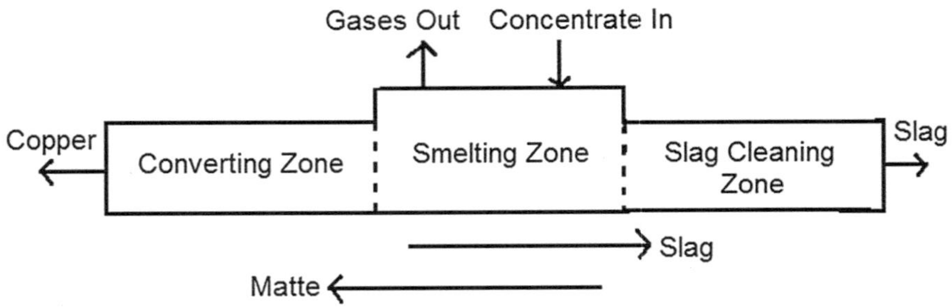

Fig. 7.8 WORCRA Reactor

(1) The process directly produces the metal, rather than a matte, from the concentrate.
(2) The heat generated by the oxidation reactions is utilized in the reactor.
(3) In the smelting and converting zones, the baths are kept in a constant motion and turbulence.
(4) The counter movement of the matte and slag takes place in the converting zone, the slag is moving under gravity.
(5) After the slag has passed through the smelting zone and the slag cleaning zone, its copper content is drastically reduced to allow it to be discarded.
(6) The gases, which contain SO_2, generated in the smelting and converting stages leave the reactor and are suitable for the production of H_2SO_4 acid.

Several factors were expected to contribute to the efficiency of the WORCRA process. First, the process injects concentrate particles which possess a large surface area. This helps in accelerating the reaction with gases in the smelting and converting zones. In the converting zone, there is a complete dispersion of the slag and the matte; whereas in conventional smelting, the reactions between the slag and the matte take place only at the slag–metal interface. Second, the counter-movement of the matte and slag leads to the effective removal of impurities such as iron. Third, in the smelting zone, valuable copper present in the slag can be reverted to the matte phase by reactions with the ferrous sulphide present in the matte. Fourth, the process maintains appropriate oxidizing and reducing conditions in different zones.

7.4.2.2 Noranda Process

The Noranda process was developed by the Noranda Mines Limited, Noranda, Canada as a continuous smelting process to go into commercial operation in 1973. The reactor, which is fuel fired at ends, produces copper metal or a high grade matte directly from the sulphide concentrate. Initially sulphide concentrate in pelletized in pelletizer, then pellets are fed into the reactor at the feed end. First pellets are melted and then oxidized by air which is coming from tuyeres. Copper or matte is collected from a tap hole of the reactor and slag is taken out from another tap hole. This slag contains high copper content; it is cooled slowly, milled and floated to get a low copper slag.

In the reactor, three layers exist together, namely, copper, matte and slag. When excess air is available, and then produces metallic copper and iron oxide, the excess air also oxidizes a part of matte to copper. This results in a decrease in the matte level and an increase in the copper level. When the amount of air is not sufficient, unoxidized iron and sulphur then tend to form matte in combination with copper. This increases the matte level and decreases in the copper level. By looking at these changes in the levels, the ratio of air to concentrate can be adjusted to yield the desired end product. The smelting rate is 4.48 tonne/day/m^3.

7.4.2.3 Mitsubishi Process

The Mitsubishi process was developed in Japan, and is already operated commercially at the Mitsubishi plant, Japan with a capacity of 650 tonne/day. Three furnaces (i.e. smelting furnace, slag cleaning furnace and converting furnace) are in a line as shown in Fig. 7.9. In the smelting furnace, concentrate is smelted to produce a matte (containing 60–65% Cu) and slag. Both the matte and the slag from the smelting furnace are transferred by gravity to the slag cleaning furnace, where they are separated from each other. The matte is transferred to the converting furnace and the slag is discarded. In the converter, the matte is oxidized to blister copper by injected oxygen-enriched air through lances from top of the bath. Limestone is added as flux to form lime ferrite slag instead of iron silicate slag. This is to ensure that a blister copper with low sulphur content is produced and to decrease copper transfer to the slag. This slag is transferred to the smelting furnace for recovery of copper. The blister copper is continuously tapped. The smelting rate is 3.41 tonne/day/m^3.

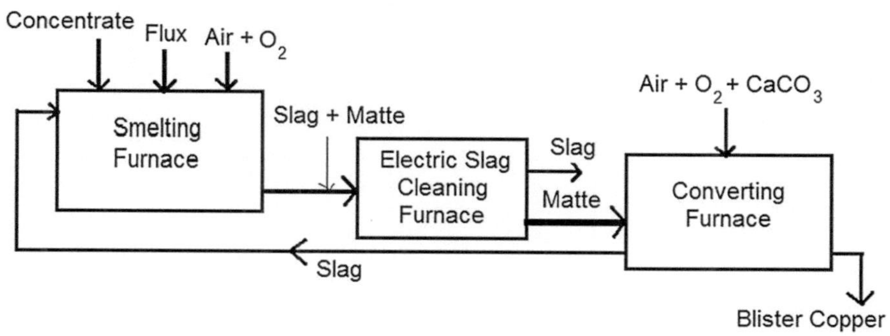

Fig. 7.9 Mitsubishi Continuous process

7.5 TORCO Segregation Process

The TORCO (treatment of refractory copper ores) process was developed to extract copper concentrate from oxidized copper ores, such as alumino-silicates, because normal leaching or flotation operations are unsuitable. Figure 7.10 shows the flow diagram of the TORCO process. In this process, grounded ore, which contains 2–6% Cu, is mixed with 0.5–2% coke or coal and 0.5–1% NaCl, then heated in a segregation reactor at a temperature of 700–800°C for about 0.5–1 h. Most of the copper gets precipitated on the carbon surface and the product consists of copper-coated carbon particles which form the segregate. The segregate and the gangue are separated from each other by flotation. The product from the reactor is continuously transferred into the flotation cell to produce a concentrate of 30–50% metallic copper. This product is sent to a smelter. The recovery of copper from this process is 85–90%.

Reactions involved in the segregation process are as follows:

(1) First of all, copper oxide is reduced to cuprous oxide by CO gas.

$$2CuO(s) + CO(g) = Cu_2O(s) + CO_2(g) \tag{7.28}$$

(2) Sodium chloride is hydrolyzed in the presence of water vapour to form HCl.

$$2NaCl(s) + H_2O(g) = Na_2O(s) + 2HCl(g) \tag{7.29}$$

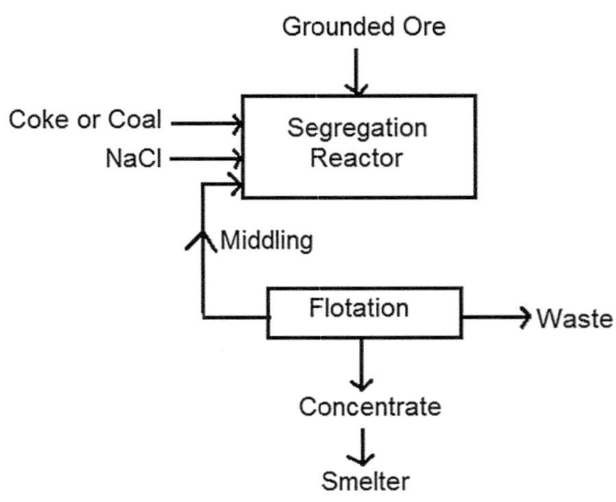

Fig. 7.10 TORCO Segregation process

(3) The copper oxide is reacted with HCl to product gaseous copper chloride.

$$Cu_2O(s) + 2HCl(g) = 2/3\,Cu_3Cl_3(g) + H_2O(g) \tag{7.30}$$

(4) The cuprous chloride and water vapours are diffused on the carbon surface, where cuprous chloride is reduced to metallic copper by hydrogen gas (H_2O vapour is reduced to H_2 gas by carbon).

$$C(s) + H_2O(g) = H_2(g) + CO(g) \tag{7.31}$$

$$2/3\,Cu_3Cl_3(g) + H_2(g) = 2Cu(s) + 2HCl(g) \tag{7.32}$$

(5) The regenerated HCl is again used for chloridization.

7.6 Recovery of Precious Metals

Recovery of precious metals from anode slime is shown in Fig. 7.11. Anode slime consist of 10–40% Cu, 2–20% Pb, 0.5–5% As, 0–4% Te, 0–25% Ni, 2–10% S, 0.5–2% Fe, 0–25% Se and 8.5–14 kg/tonne of Ag, 0.25–1.7 kg/tonne of Au. To remove Cu, the slime is first leached with dilute sulphuric acid and residue is treated again by sulphuric acid at 200°C to get all Cu in solution. The copper solution contains tellurium which is recovery by cementation with active copper. After removal of copper, the slime is roasted at 700°C in air, selenium oxide is formed from which selenium is recovered. The selenium free slime is then smelted with flux and passing air to make lead slag and metals consist of tellurium, silver, gold platinum and palladium. These metals is treated with soda ash, silica and borax ($Na_2B_4O_7$) to get silver rich alloy (> 90% Ag). The alloy is cast in the form of an anode and electrolyzed in the silver nitrate bath to recover silver and anode slime containing Au, Pt and Pd metals.

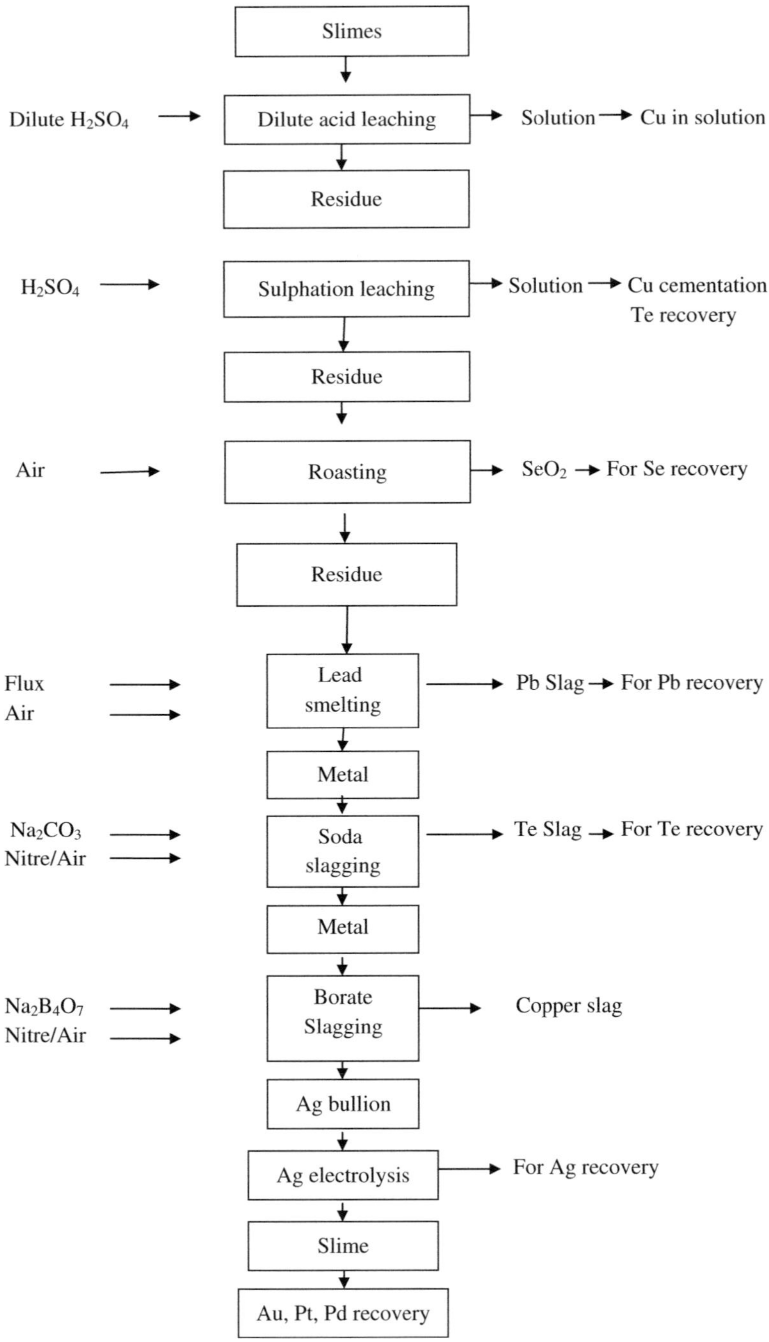

Fig. 7.11 Flow sheet for recovery of precious metals

7.7 Hydrometallurgical Process of Copper

About 80–85% of the world's total production of copper is done by the pyrometallurgical process, rest is by the hydrometallurgical process. The hydrometallurgical process is mainly used in the extraction of copper from oxidized ores or low grade sulphide ores.

7.7.1 *Ferric Chloride Leaching*

It is well known that chalcopyrite mineral is difficult to leach. However, it is found that ferric chloride can be used as leaching reagent for chalcopyrite. The reaction between chalcopyrite and ferric chloride is as follows:

$$CuFeS_2 + 3FeCl_3 = CuCl + 4FeCl_2 + 2S \tag{7.33}$$

This reaction is not only dissolved copper but also produces elemental sulphur. This is a distinct advantage over the pyrometallurgical process which emits SO_2 gas.

Figure 7.12 shows the flow sheet for hydrometallurgical process of copper extraction. Copper is recovered by cementation and fire refining. The ferric chloride is regenerated by the chlorination of the ferrous chloride solution which is a by-product of the process.

The most extensively applied solvent for oxidized copper ore is dilute sulphuric acid, for not only does it readily dissolve the metal but also renders it in a form suitable for subsequent precipitation and recovery. A disadvantage, however, is that, in addition to the oxidized copper minerals, the gangue, when composed of such compounds as carbonate, silicate and oxide is also attacked. These dissolved impurities are objectionable not only from the viewpoint of the amount of acid they consume, but also because they tend to accumulate in the solution when recycled.

Ferric sulphate has been used in copper leaching for in addition to oxidized copper ore it also acts as a solvent for sulphides.

$$Cu_2S + Fe_2(SO_4)_3 = 2CuSO_4 + 2FeSO_4 \tag{7.34}$$

7.7.2 *Leaching of Low Grade Ores*

Hydrometallurgical process of copper extraction is the best suited for low grade copper ore (Fig. 7.13). The low grade ore is accumulated in the form of a heap and dilute acid is percolated through it. After percolation, the leach solution is either

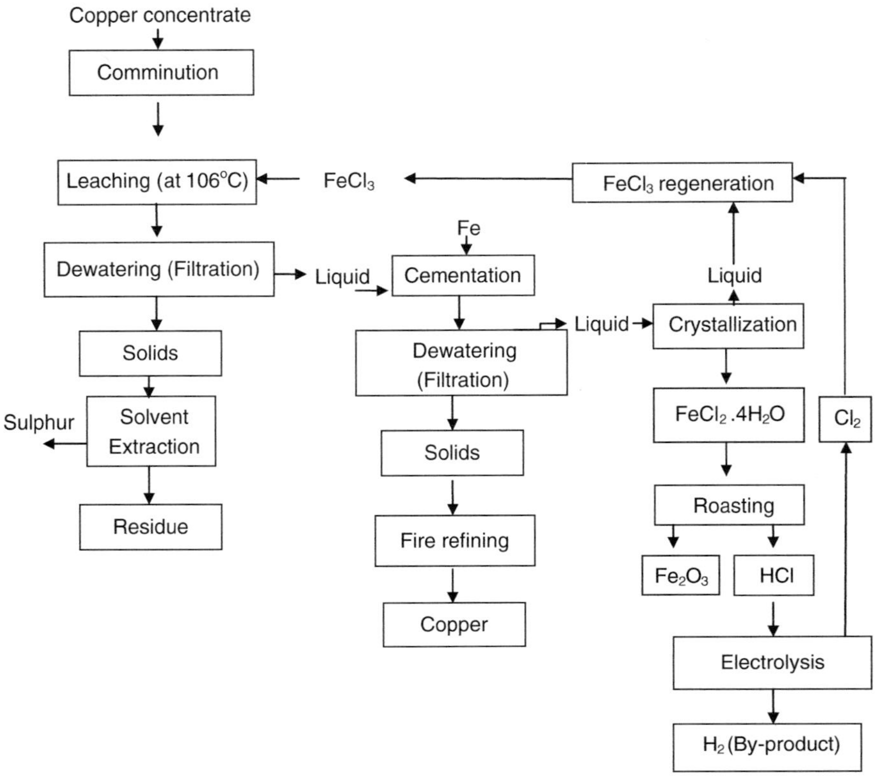

Fig. 7.12 Flow sheet for hydrometallurgical process of copper

treated with scrap iron to recover cement copper or subjected to solvent extraction to extract copper and also to provide a solution suitable for electro-winning. To accelerate the leaching rate bacteria can be used.

In case of an oxidized copper ore, leaching is carried out by using a spent copper electro-winning electrolyte that contains about 100 g/l H_2SO_4. The resultant leach liquor that contains about 50 g/l copper is used for electro-winning metallic copper.

7.7.3 Leaching of Roasted Sulphide Concentrates

A sulphide concentrate of copper is roasted prior to leaching. In a fluid bed roaster, a roasting temperature higher than 650°C is required to achieve good roasting kineties. During the subsequent electro-winning of the $CuSO_4$ solution (obtained from the dilute acid leaching of the roasted concentrate), H_2SO_4 is generated. The reaction is

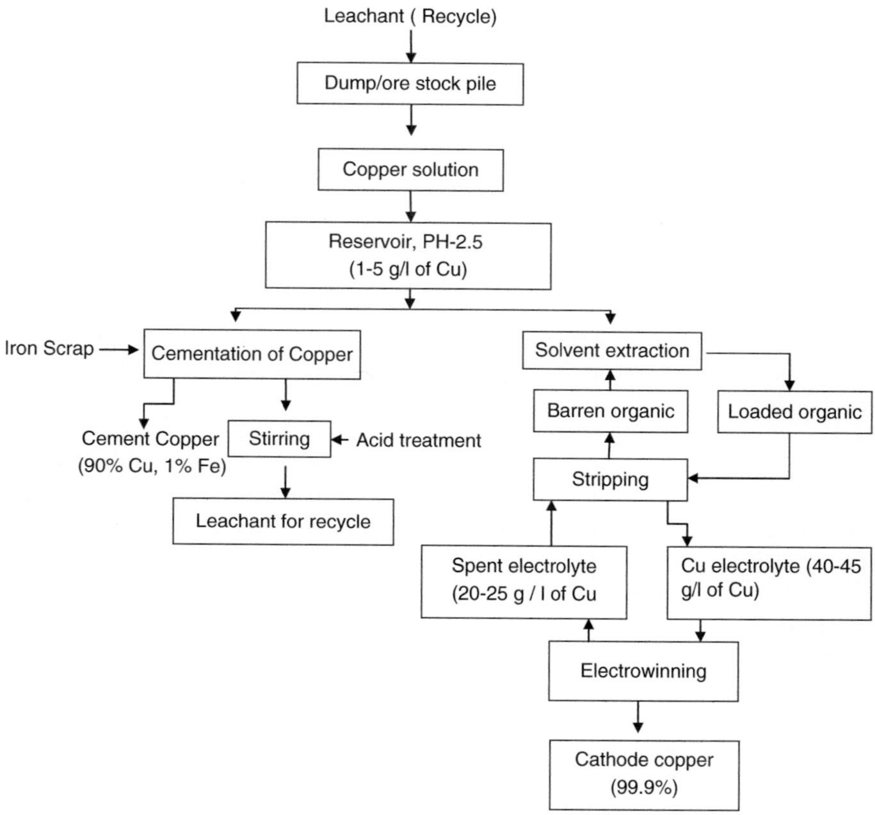

Fig. 7.13 Flow sheet of leaching for low grade copper ore

$$Cu^{2+} + SO_4^{2-} + H_2O = Cu + H_2SO_4 + 1/2\,O_2. \tag{7.35}$$

The sulphuric acid in the spent electrolyte is normally used for the leaching of copper oxide ores.

7.8 Properties

Atomic weight of copper is 63.54. Specific gravity of copper is 8.96. Melting point and boiling point are 1083°C and 2580°C respectively. At metallurgical processing temperatures the pressure of vapours of copper, its oxides and sulphidies are so small that they may be considered practically non-volatile. Copper and its alloys have a unique combination of properties, such as, high electrical and thermal conductivities, adequate mechanical properties, and high resistance to various types of corrosion. The properties of copper differ depending on the nature of the impurity present in it. Oxygen adversely affects the electrical properties of copper. When present in very small quantity, oxygen (0.02–0.04%) is not harmful; in fact, it actually increases the electrical conductivity by removing other impurities from the solid solution. Impurities such as sulphur, selenium, bismuth, antimony, and tellurium have harmful effects on both the mechanical and electrical properties of copper and have to be removed during refining. Annealed copper when work-hardened loses about 3% of its conductivity; after re-annealing, however, restores it. The electrical conductivity of copper is affected to a great extent by the addition of other elements; phosphorus, silicon and arsenic are highly decreasing, whereas silver, cadmium, zinc and oxygen (in small amounts) have little influence on the electrical conductivity of copper.

Copper is next to silver in respect of its thermal conductivity. The effect of temperature on the thermal conductivity of copper is very little. By rising to 200°C, the loss of thermal conductivity is less than 1%. The effect of impurities leads to a reduction of thermal conductivity in approximately the same degree as its electrical conductivity. The thermal conductivities and electrical resistivities of silver, copper, gold, and aluminium (as shown in Table 7.1) are outstanding when compared with those of other metals. Since silver and gold are highly expensive, they cannot be used on a commercial scale for electrical applications. In this area, copper has been found ideal.

By location in the Periodic table (Fig. 1.1), close to copper are metals of the same subgroup like gold and silver. They do not oxidize in air and occur naturally in native state. Molten copper, gold and silver are readily soluble in one another.

Oxygen-free copper is in demand for welding and brazing, in addition to operations involving heavy degrees of cold work such as deep drawing. Deoxidized copper contains up to 0.04% phosphorus, the presence of which has an adverse effect on the electrical conductivity, diminishing it by some 20%. Arsenic (up to 0.5%) is added to copper with the intent of obtaining increased strength and toughness, also raising the softening temperature by about 100°C.

Under severe conditions of atmospheric corrosion, copper is extremely resistant to attack, as a result of the formation of a film of oxide. When exposed to the action

Table 7.1 Selected physical properties of some metals

Metal	Specific gravity	Melting point (°C)	Thermal conductivity at 20°C (cal cm^{-1} s^{-1}°C^{-1})	Electrical resistivity at 20°C (10^{-6} Ω cm)	Relative electrical conductivity (copper = 100)
Silver (Ag)	10.49	960	1.00	1.59	106
Copper (Cu)	8.94	1083	0.94	1.67	100
Gold (Au)	19.32	1063	0.71	2.19	72
Aluminium (Al)	2.7	660	0.53	2.66	62

of air or water for long periods, this oxide skin is converted to sulphides or to complex basic sulphates and carbonates. Copper is excellent resistance to most acids and alkalis, other than the strong mineral acids and ammonia.

Since copper is ductile in nature, it may readily be worked cold by any process involving rolling, wire-drawing, spinning and pressing etc. Cold working increases both strength and hardness but decreases elongation. If the metal is required in the soft condition after cold working this may be achieved by annealing at 400–600°C. Due to its exceptional plasticity, copper can be rolled, extruded or forged over a wide range of temperatures. Tensile strength of cast copper is about 15.75 to 17.325 kg/mm^2. Rolling, forging or other types of hot and cold working followed by annealing raises its tensile strength to upwards of 22.05 kg/mm^2 with about 45% elongation. Izod impact values of most commercial copper in the fully annealed condition vary between 54.24 and 67.8 J.

7.9 Applications

Copper is the most important non-ferrous metal. It can be used in alloy as well as unalloyed state; the world output of this metal is greater than any other non-ferrous metal. The addition of a metal or metals to copper forms alloys which have superior hardness, mechanical and other special properties, and they are in wide use for many applications. Copper can make alloys with Zn, Sn, Be, and Ni. Pure copper is used extensively in electrical applications due to its very good electrical conductivity. Nowadays, aluminium is gradually replacing copper in the power transmission lines and in other electrical applications.

In non-electrical applications, copper is used in the form of an alloy. Some of the commonly used alloys are Cu–Zn brasses, Cu–Sn bronzes and Cu–Ni–Zn alloys. Cu–Zn brasses that contain up to 36% Zn are known as α-brasses, and have exceptional cold working properties. Cu–Zn brasses, which contain between 37 and 45% Zn, are known as α-β-brasses; those which contain more than 45% Zn are known as β-brasses. α-β-brasses are suited for hot working, which improves as the zinc content increases, reaching a maximum at about 40%.

Cu–Sn bronzes, which have a higher mechanical strength than metallic Cu, and fatigue-resistant and corrosion-resistant properties. Cu–Ni alloys containing 50–70% Cu have excellent ductility and offer a high resistance to corrosion.

Cu–Al bronzes form a range of alloys based on copper and aluminium with additions of iron, nickel and manganese. The protective film of aluminium oxide formed on the alloys by heating. Cu–Al bronzes are among the best of the copper base alloys for service at moderately elevated temperatures. They show good resistance to all normal forms of corrosion. Aluminium content for normal cold work is about 8%; 10% Al–Cu bronze can be fabricated in hot conditions and can be used for casting.

Addition of silicon to copper (silicon bronze) not only increases the strength and gives good resistance to corrosion, especially by acids, but also improves the weldability since silicon is a strong deoxidizer and helps to maintain the molten weld metal in a deoxidized condition. Addition of silicon to brass enhances the mechanical properties and increases the fluidity of the molten metal for die castings. The most promising alloy contains 15% Zn, 4% Si, and the rest Cu. This bronze gives good clean castings and can be easily extruded, hot or cold worked.

The copper manganese aluminium alloys are ductile and malleable and certain of these alloys posses a low temperature coefficient of electrical resistance. Thus, they (85% Cu, 13% Mn and 2% Al) are particularly suitable for the fixed resistances of electrical instruments. Although manganese is non-magnetic, alloy containing 65% Cu, 25% Mn and 10% Al can be magnetized to form permanent magnets. This material is relatively soft and ductile and can be formed by bending and machining into any shapes which would be difficult for magnetic steels.

An alloy containing 2% beryllium, after suitable heat treatment, is the hardest and strongest of all copper base alloys, and is used for producing tools. The resistance to fatigue and corrosion shown by the alloy is responsible for its main outlet, namely spring material, bushings and bearings.

Chapter 8
Aluminium

8.1 Introduction

Aluminium (Al) is a lighter metal than copper, zinc and lead. It was discovered by Davy in 1805. Wohler isolated the metal by the reduction of aluminium chloride with potassium 22 years later. In 1856 Deville in France substituted potassium by sodium as reducing agent. In 1886, C. Hall in the USA and P. Heroult in France suggested obtaining aluminium by electrolysis of alumina dissolved in molten cryolite. This Hall-Heroult method remains a most convenient and economic production method of aluminium which is popular throughout the world. Aluminium has good electrical and thermal conductivities, and resistance to atmospheric corrosion. Aluminium (in oxide form) is contained in 15% of the Earth's crust.

8.2 Sources

The most common aluminium minerals are alumina trihydrates as gibbsite ($Al_2O_3.3H_2O$) and alumina monohydrates as diaspore ($Al_2O_3.H_2O$). The theoretical Al_2O_3 present in gibbsite is 65.4%, and in diaspore is 85.4%. Bauxite is a mixture of these two minerals and contains varying amounts of impurities such as Fe_2O_3, TiO_2 and SiO_2. The bauxite ores found in India normally contain 58–67% alumina. These ores have a high content of TiO_2 (5–10%) and a low content of silica (0.5–2.0%).

S.K. Dutta and D.R. Lodhari, *Extraction of Nuclear and Non-ferrous Metals*,
Topics in Mining, Metallurgy and Materials Engineering,
https://doi.org/10.1007/978-981-10-5172-2_8

8.3 Extraction of Aluminium

The aluminium oxide is very stable. The carbothermic reduction of aluminium oxide has not been adopted because:

(1) It requires a very high temperature (above 2000°C),
(2) At that temperature, aluminium carbide is formed,
(3) The refractory required, for such high temperature, is readily available and very expensive.

The extraction of aluminium by the electrolysis of aluminium salts in an aqueous medium is not feasible, due to decomposition of the aqueous solution which results in the evolution of hydrogen at a much lower voltage than that needed for the electro-deposition of aluminium ions.

Hence, the universally adopted process for the extraction of aluminium is the Hall-Heroult process, in which alumina (Al_2O_3) dissolved in cryolite ($3NaF.AlF_3$) is electrolyzed. The alumina, obtained from the Bayer process, is a pure form to be used as the electrolyte in the electrolytic cell. But this process consumes a large amount of electrical energy.

8.3.1 Bayer Process

In the Bayer process, high purity alumina is produced from bauxite (Fig. 8.1). The bauxite is leached by sodium hydroxide under high pressure (25 atmospheres) and a temperature at 220°C to form soluble sodium aluminate, from which $Al(OH)_3$ is precipitated. The alumina is obtained by calcinations of $Al(OH)_3$.

Initially, bauxite obtained from the mines is crushed and ground to a very fine size to separate silica from bauxite, since silica is concentrated in the finer fractions. By comminuting, alumina can be enriched in the coarser fractions. The coarser fractions are further ground along with caustic soda in a ball mill to form slurry, which is fed into an autoclave (at 220°C temperature and 25 atmospheres pressure). The alumina, in the bauxite, is dissolved in caustic soda:

$$Al_2O_3.H_2O + 2NaOH = 2NaAlO_2 + 2H_2O \tag{8.1}$$

$$Al_2O_3.3H_2O + 2NaOH = 2NaAlO_2 + 4H_2O \tag{8.2}$$

The impurities, present in bauxite, leave as suspended solids in the solution. The dissolution of the alumina monohydrate [reaction (8.1)] requires a temperature of 230°C and a time of 3 h, whereas that for the alumina trihydrate [reaction (8.2)] requires a temperature of 180°C and a time of 1 h. Since the bauxite is a mixture of the monohydrate and the trihydrate, the temperature used is 180–220°C and the time allowed is 2.5 h. Under these conditions, 86–88% alumina in the bauxite goes

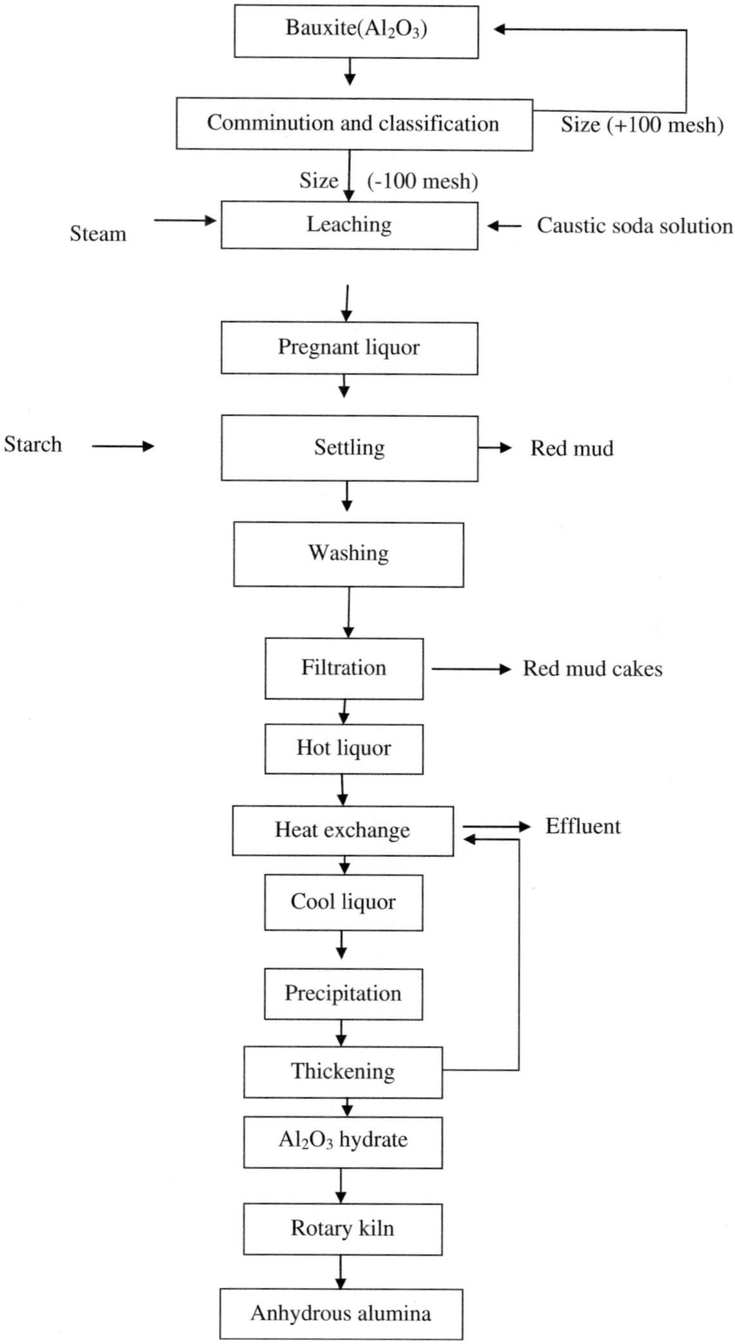

Fig. 8.1 Flow sheet of Bayer process

into the solution. At such high temperatures, a sufficiently high pressure (5–25 atmospheres) must be applied in order to retain water in the liquid state.

Then the liquid is cooled below 100°C, and the settling of red mud is speeded up by the addition of starch. Any remaining red mud in the liquid is eliminated by washing with hot water. Finally the liquid is filtered; the filtrate obtained is a clear solution of sodium aluminate. The residue left behind after filtration consists mainly of ferric hydroxide, silica, and non-dissolved alumina.

The heat of the hot liquid is extracted by heat exchanger for conservation and for cooling the liquid below the critical temperature required for alumina precipitation. Very fine, freshly prepared aluminium hydroxide is added to create nuclei in order to accelerate the precipitation of $Al(OH)_3$. The precipitate is separated from the liquid in the thickener, and the product is calcined in a rotary kiln to produce anhydrous alumina. Depending on calcinations temperature, resultant Al_2O_3 is either a loose hygroscopic γ-Al_2O_3 (800°C) or a more dense α-Al_2O_3 (1200°C), but generally the product is a mixture of both components due to flow rate of transformation.

$$2NaAlO_2 + 4H_2O = 2Al(OH)_3 + 2NaOH \tag{8.3}$$

$$2Al(OH)_3 = Al_2O_3 + 3H_2O \tag{8.4}$$

The effect of various factors on the efficiency of the Bayer process is as follows:

(1) The finer the bauxite, the better dissolution of alumina in the leach liquid.
(2) Wet grinding is more efficient than dry grinding, and also cuts down the time required for dissolution.
(3) Dissolution is accelerated by increasing temperatures; above 100°C water would be evaporated, hence high pressures up to 25 atmospheres are applied to attain a temperature up to 220°C.
(4) There is a lower limit to the temperatures. Below that limit, aluminium hydroxide may precipitate, leading to a loss of alumina.
(5) The sensible heat, collected from hot liquid, is used for producing steam for dissolution.
(6) The entire amount of alumina in solution is not allowed to precipitate because this may lead to the simultaneous precipitation of dissolved silica. Therefore, the precipitation is deliberately kept incomplete; the residual solution is recirculated to the dissolution stage.
(7) For the efficient calcinations of alumina, the rotary kiln should be able to attain a temperature as high as 1400°C. The decomposition of $Al(OH)_3$ produces $Al_2O_3.H_2O$ and $Al_2O_3.3H_2O$ (water is chemically bound). The minimum temperatures required for the dehydration of the monohydrate and the trihydrate are 1200°C and 800°C respectively.

8.3.2 Hall-Heroult Process

In 1886, Hall in the USA and Heroult in France independently patented a process for the commercial production of aluminium. This process is based on the electrolytic decomposition of alumina dissolved in a liquid bath of cryolite. A source of cheap electric power is the primary requisite for the Hall-Heroult process, which also requires high purity alumina, cryolite, and ashless carbon electrodes.

The electrolysis of aluminium from alumina is done in an electrolytic cell (Fig. 8.2). The cell contains a rectangular refractory lined steel container which is 5 m long, 2 m wide and 1 m deep. The bottom of the container is conducting and acts as cathode which consists of refractory bricks of carbon bonded by tar binder. Iron plate and iron bar are embedded in the carbon mixed for good conduction to form cathode. Anodes are one of two forms: firstly, the anode may be prebaked carbon electrodes that are replaced from time to time. Embedded in the anode is a copper bar for contact with the cell bus bar. Up to 24 anodes form the complement of one cell. The average life is 8–10 days. Secondly, the continuous electrodes by continuous feeding carbonaceous paste (hot mixture of 70% petroleum or pitch coke and 30% pitch as binder) into the steel moulds. The carbonaceous paste is baked in the cell itself during electrolysis by the heat resulting from the resistance offered to the passage of the electric current, and acts as the anode. The baked carbon anode and the lining cathode must be sufficiently strong (250–300 kg/cm^2 of compressive strength) and dense (< 25% porosity). Electrical contact is established by rows of iron pins attached to the casing and to aluminium bus bars which also help to carry the weight of the electrode. As the electrode is consumed and lowered into the electrolyte the sheet aluminium casing melts joints the electrolyzed metal in the bottom of the cell.

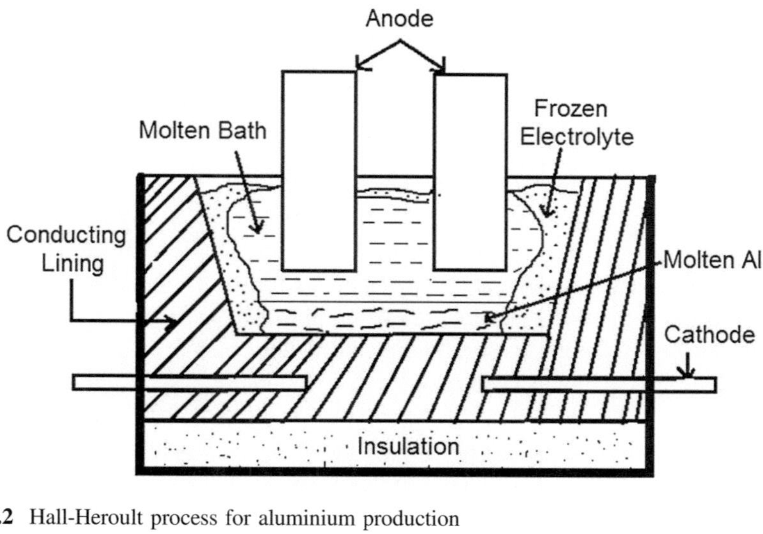

Fig. 8.2 Hall-Heroult process for aluminium production

A tap hole for the collection of molten aluminium metal is located on the bottom of the furnace. The density of the bath must be less than that of liquid aluminium, or else the liberated metal will float and upset cell operation. Molten aluminium is collected at the bottom, as its density (2340 kg/m^3) is higher than that of the electrolyte (2030 kg/m^3).

Since alumina is not an ionic compound, very few ionic melts are capable of dissolving it to any appreciable extent. However, cryolite ($3NaF.AlF_3$), a naturally occurring mineral, can dissolve up to 15% alumina at 1000°C. Cryolite has been successfully synthesized by passing hydrogen fluoride (HF) through sodium aluminate ($Na_6Al_2O_6$) solution, The hydrogen fluoride is produced from naturally occurring CaF_2.

$$CaF_2(s) + (H_2SO_4)_{aq} = 2(HF)_{aq} + CaSO_4(s) \tag{8.5}$$

$$(Na_6Al_2O_6)_{aq} + 12(HF)_{aq} = 2Na_3AlF_6(s) + 6(H_2O)_{aq} \tag{8.6}$$

Small amounts of CaF_2 and NaF are added to the cryolite bath to enhance its conductivity and to improve the metal recovery. Cryolite melts at 990°C, at which temperature it will take up to 10% of its weight of alumina. At 1000°C, the density of the molten cryolite is approximately 2100 kg/m^3 and that of Al_2O_3 is 3960 kg/m^3, which sink at the bottom of the cell. When the bath containing about 10% CaF_2, 80% cryolite, 5% Al_2O_3, and a small amount of NaF is electrolyzed, the aluminium (density 2340 kg/m^3) is liberated which is heavier than the bath and collected at the bottom of the bath. In this way, the aluminium metal is protected against atmospheric oxidation. The density is one of the factors which limit the concentration of Al_2O_3 in the bath; the higher the amount of Al_2O_3 in the bath, the higher its density and consequently the sinking of aluminium is rendered more difficult. Depth of electrolyte is maintained at about 0.25 m.

During operation, the cell is filled with cryolite and the anode is lowered into that. Current is passed through the cell until the cryolite (melting point 990°C) melts. When the bath attains the molten state, then only alumina is charged to the cell. The alumina decomposes to alunimium and oxygen, which is an endothermic reaction, necessary heat is supplied by the resistance offered by both the electrodes and the electrolyte.

$$2Al_2O_3 = 4Al + 3O_2 \quad \Delta H^\circ = 838,683\,\text{J/mol of}\,Al_2O_3 \tag{8.7}$$

This oxygen reacts with carbon in anode to form CO and CO_2 gases. The agitation of the bath takes place due to evolution of CO and CO_2 gases; as well as magnetic stirring due to the current flowing through the bath from anode to cathode. The agitation help maintain fresh additions of alumina in suspension long enough for dissolution in the electrolyte. If the alumina content of the bath (normally 5–10%) decreases below 2%, then normal contact between the anode and the bath is interrupted by a gas film which increases the resistance which causes the voltage to rise from 5–6 V to as much as 40–50 V. Fluorine gas is deposited at the anode surface where it forms a gas film which is electrically insulating, occasioning a rise in the voltage. The current then is carried by arcing between the carbon anode and the electrolyte. As a result, normal operation is hampered. This effect is known as the anode effect. With addition of sufficient alumina to the bath the voltage returns to its normal value.

When alumina is dissolved in molten cryolite, the ions are present as follows:

In solute: Al^{3+}, O^{2-} .
In solvent: Na^{+}, Al^{3+}, AlF_6^{3-}, F^{-}.

The oxygen ions (O^{2-}) are the only ions foreign to the solvent and may react with Al^{3+} to form AlO^{+} and AlO_2^{-} in the melt. Moreover, Al_2O_3 may go into solution in the form of AlO^{+} and AlO_2^{-}. During electrolysis, several reactions involving all the ions take place simultaneously in the melt.

$$Na_3AlF_6 = NaAlF_4 + 2NaF \tag{8.8}$$

$$NaAlF_4 = NaF + AlF_3 \tag{8.9}$$

The NaF produced initiates a number of reactions as indicated by the following two theories.

Theory 1: The NaF is dissociated into Na and F:

$$6NaF = 6Na + 6F \tag{8.10}$$

$$6Na + 2AlF_3 = 6NaF + 2Al \tag{8.11}$$

$$6F + 2Al_2O_3 = 2AlF_3 + 3O_2 \tag{8.12}$$

$$3O_2 + 6C = 6CO \quad \text{or } 3O_2 + 3C = 3CO_2 \tag{8.13}$$

Theory 2: The NaF reacts with Al_2O_3 to form Na_2O:

$$6NaF + Al_2O_3 = 2AlF_3 + 3Na_2O \tag{8.14}$$

$$3Na_2O = 6Na + 3/2\,O_2 \tag{8.15}$$

$$6Na + 2AlF_3 = 6NaF + 2Al \tag{8.16}$$

$$3/2\,O_2 + 3C = 3CO \text{ or } 3/2\,O_2 + 3/2\,C = 3/2\,CO_2 \tag{8.17}$$

The foregoing two theories attribute the constant presence of sodium vapour and fluorine above the cryolite bath to the decomposition of NaF. It is also observed that the depletion of alumina or the lowering of its activity enhances the formation of fluorine [reaction (8.10)] or NaF [reactions (8.9) and (8.11)]. An increased generation of fluorine causes the anode effect, which prevents normal operation.

8.3.2.1 Decomposition Potential

To estimate the voltage required for the process, first decomposition reaction may be written as:

$$1/2\,Al_2O_3(\text{solution in}\,Na_3AlF_6) = Al(l) + 3/4\,O_2(g), \quad \Delta G_{8.18} \tag{8.18}$$

This reaction (Eq 8.18) may be split up into a series of reactions, for which the standard free energies are known, as follows:

$$Al(l) + 3/4\,O_2(g) = 1/2\,Al_2O_3(s), \quad \Delta G_{8.19} \tag{8.19}$$

$$1/2\,Al_2O_3(s) = 1/2\,Al_2O_3(l), \quad \Delta G_{8.20} \tag{8.20}$$

$$1/2\,Al_2O_3(l) + Na_3AlF_6(l) = 1/2\,Al_2O_3(\text{solution in}\,Na_3AlF_6), \Delta G_{8.21} \tag{8.21}$$

At 1000°C, for the reaction:

$$4/3\,Al(l) + O_2(g) = 2/3\,Al_2O_3(s), \Delta G_{8.22} \tag{8.22}$$

The free energy change of reaction (8.22), $\Delta G_{8.22} = -\ 861.9$ kJ
Hence, $\Delta G_{8.19}$ is [(3/4) × (−861.9)] = − 646.43 kJ /mol of Al.
$\Delta G_{8.20}$ can be calculated from the heat of fusion for Al_2O_3. $\Delta G_{8.19}$ is equivalent to converting ½Al_2O_3, at 1000°C, to the molten state. From standard book's data*, $\Delta G_{8.20}$ can be obtained as 15,899.2 J. $\Delta G_{8.21}$ is the free energy change for the dissolution of Al_2O_3 into cryolite.
Therefore, for 1 mol of Al_2O_3,

$$\begin{aligned}\Delta G_{8.21} &= -RT\ln k = -RT\,\ln\left[a_{Al2O3\,\text{in}\,Cry}/a_{Al2O3}\text{x}\ a_{Na3AlF6}\right]\\ &= -RT\,\ln\left[a_{Al2O3\,\text{in}\,Cry}\right]\end{aligned} \tag{8.23}$$

Since a_{Al2O3} and $a_{Na3AlF6}$ are unity for pure states.
If the bath contains 8 wt% alumina and rest cryolite, then the mole fraction of alumina can be calculated as follows:
Mole fraction of

$$\begin{aligned}Al_2O_3 &= [(\text{wt}\%Al_2O_3/\text{mol wt of}\,Al_2O_3)/\{(\text{wt}\%\,Al_2O_3/\text{mol wt of}\,Al_2O_3)\\ &\quad + (\text{wt}\%\,\text{cryolite}/\text{mol wt of cryolite})\}]\\ &= [(8/102)/\{(8/102) + (92/210)\}]\\ &= [0.078/(0.078 + 0.438)] = 0.15.\end{aligned} \tag{8.24}$$

Assuming that the solution is ideal and the activity of alumina is equal to its mole fraction.
Therefore, $\Delta G_{8.21} = -\ RT\ \ln[a_{Al2O3\ \text{in}\ Cry}] = -\ [(8.314 \times 1273)\ \ln(0.15)] = 20{,}078.4$ J

Since R = 8.314 J/mol.K and T = (1000 + 273) K = 1273 K
Now,

$$\Delta G_{8.18} = -(\Delta G_{8.19} + \Delta G_{8.20} + \Delta G_{8.21}) = -[(-646430) + 15899.2 + 20,078.4)] \\ = -(-610452.4) = 610,452.4\,\text{J} \tag{8.25}$$

We know,

$$\Delta G = FzE \tag{8.26}$$

where F is the Faraday's constant = 96,500 J/v. g.equivalent; z is the valancy of aluminium and E is the decomposition potential for reaction (8.18).

Hence

$$\Delta G_{8.18} = 610,452.4\,\text{J} = FzE = 96500\,zE \tag{8.27}$$

Therefore,

$$E = [610452.4/(96500 \times 3)] = 2.11\,\text{v} \tag{8.28}$$

This theoretically calculated value compares well with the experimentally obtained values between aluminium and electrodes, which lies between 2.1 and 2.15 v.

In practice, a consumable graphite electrode serves as the anode. To calculate the decomposition potential of an aluminium (liquid) cathode and carbon (solid) anode, the reaction (8.18) has to be combined with the carbon oxidation reactions:

$$3/4\,O_2(g) + 3/4\,C(s) = 3/4\,CO_2(g), \Delta G_{8.29} \tag{8.29}$$

$$3/4\,O_2(g) + 3/2\,C(s) = 3/2\,CO(g), \Delta G_{8.30} \tag{8.30}$$

From standard book*, the thermodynamics values of $\Delta G_{8.29}$ and $\Delta G_{8.30}$ obtain as −296,401.59 J and −334,936.87 J respectively, at 1000°C. Combining reaction (8.29) with reaction (8.18):

$$1/2\,Al_2O_3(\text{solution in } Na_3AlF_6) + 3/4\,C(s) = Al(l) + 3/4\,CO_2(g) \quad \Delta G_{8.31} \tag{8.31}$$

Therefore,

$$\Delta G_{8.31} = \Delta G_{8.18} + \Delta G_{8.29} = 610,452.4\,\text{J} + (-296,401.59\,\text{J}) = 314,050.81\,\text{J} \\ = 96,500\,zE$$

So, E = 1.08 v.

Now, Combining reaction (8.30) with reaction (8.18):

$$1/2\,Al_2O_3(\text{solution in } Na_3AlF_6) + 3/2\,C(s) = Al(l) + 3/2\,CO(g) \quad \Delta G_{8.32} \tag{8.32}$$

Therefore,

$$\begin{aligned}\Delta G_{8.32} &= \Delta G_{8.18} + \Delta G_{8.30} = 610,452.4\,J + (-334,936.87\,J) = 275,515.53\,J \\ &= 96,500\,zE\end{aligned}$$

So, E = 0.95 v.

From the above calculation, it is evident that graphite aids the decomposition of alumina because it reacts with the oxygen to form CO and CO_2. Under normal operating conditions, about 50% by volume of CO is produced; the amount may increase considerably at a low current density. In general the values of E calculated from reactions (8.31) and (8.32) are of equal importance. Hence, the average theoretical decomposition potential is taken, i.e. [(1.08 + 0.95)/2] or 1.015 v.

8.3.2.2 Factors Affecting Electrolysis

The certain operating factors which affect the electrolysis of alumina are as follows:

(1) The bath temperature is an important factor. As a rule, an increase in temperature decreases the current efficiency. For every 4°C rise in the bath temperature (above the normal 970°C), the current efficiency falls by 1% because the higher the temperature, the more the number of side reactions and the greater the dissolution of the metal in the bath.

(2) The higher the current density, the greater the current efficiency. Therefore, the current density is maintained at a very high value.

(3) The lower the density of the bath, the higher the rate of separation of the metal liberated from the bath. The density can be lowered by using a high AlF_3/NaF ratio; a ratio of slightly less than 3 is considered optimum. When the quantity of AlF_3 is higher than NaF; the conductivity of the bath is decreased, the solubility of alumina is reduced, and a large volume of fluorine is produced. On the other hand, a high NaF content reduces the current efficiency due to an increase in the concentration of sodium ions and the deposition of sodium on the cathode. The sodium deposition directly lowers the current efficiency.

(4) The current efficiency is also affected by the interpolar distance, which is the distance from the bottom of the anode to the top of the molten bath. The current efficiency increases along with the interpolar distance, reaching a maximum (about 90%) when the distance is about 0.65 m.

(5) The cell efficiency can be improved by strictly controlling the addition of alumina. The current efficiency is a minimum when the Al_2O_3 content in the bath is 4%. At values lower and higher than 4%, the current efficiency increases.

The current efficiency (CE) has an empirical relationship with various operating variables as follows:

$$\begin{aligned}CE = {} & 254.92 - 1.7C_{Al2O3} + 0.39C_{Al2O3}^2 + 0.45AlF_3 + 0.055X_{AlF3}^2 + 0.3C_{CaF2} \\ & - 0.23T + 129/(D+1) + 25i\end{aligned} \tag{8.33}$$

where T is the temperature of the cell (°C), i is the anode current density (A/cm^2), D is the inter-electrode distance (cm), C_{Al2O3} is the wt% Al_2O_3, C_{CaF2} is the wt% of CaF_2, and X_{AlF3} is the mol fraction of AlF_3.

8.3.3 Refining of Aluminium

The purity of aluminium metal produced by the Hall-Heroult process is about 99.5%, this is sufficient for most alloying purposes. However, for electrical applications and for canning, a higher degree of purity is required. To achieve this higher purity, an electrolytic method is used which is also known as the three layer process (Fig. 8.3). The graphite electrode is used as cathode and carbon bottom acts as anode. Electrolytic refining takes place in a cell in which the anode is at the bottom, the current travelling up through the bath and being collected at the top by graphite electrodes. This method depends for its operation on the adjustment of the specific gravity of the electrolyte so that it is lighter than an anodic copper-aluminium alloy, but heavier than pure aluminium. A dense electrolyte (which consists of 36% aluminium fluoride, 30% cryolite, 18% barium fluoride, and 16% calcium fluoride) forms the middle layer (density is 2800 kg/m^3). The impure metal is alloyed with copper (28–30% Cu), and this heavy alloy (density is 4500 kg/m^3) forms the bottom layer. The purified aluminium (99.99%, density is 2300 kg/m^3) is floated upward during electrolysis to form the top layer. The temperature of electrolysis is at about 950°C. Magnesite refractory acts as insulating wall.

Refining reactions involved as follows:

$$\text{At anode}: Al \rightarrow Al^{3+} + 3e \tag{8.34}$$

$$\text{At cathode}: Al^{3+} + 3e \rightarrow Al \tag{8.35}$$

Impure aluminium is dissolved at the anode to form Al^{3+} ions; then aluminium ions (i.e. Al^{3+}) are deposited at the cathode in the pure form.

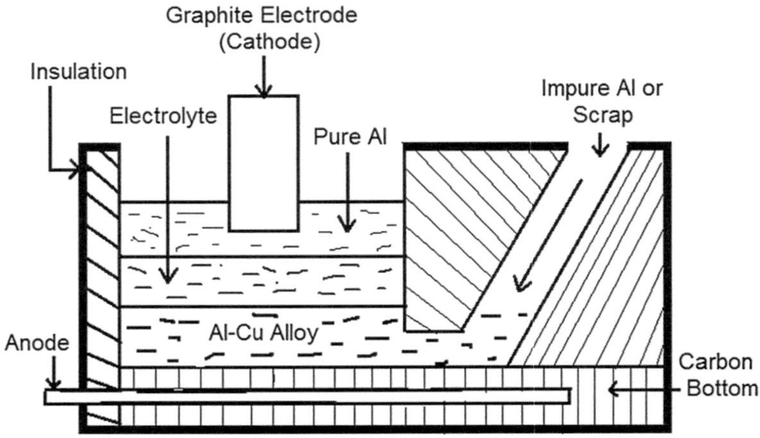

Fig. 8.3 Refining of aluminium

Commercial aluminium is poured in a molten condition into the well connected with the anode alloy metal in the bottom of the cell. Current from this layer causes pure alunimium to pass through the electrolyte and float on the top layer from where it is periodically taken out. The impurities which are retained in the bottom alloy layer gradually build up and having a relatively low solubility crystallize out in the later stage. These impurities consist mainly of iron, silicon, zinc and copper entangled with aluminium. The consumption of the electrolyte and graphite electrode each is equivalent to about 6% of the aluminium metal produced.

This method is also applicable for the refining of aluminium scrap without magnesium. This process will be ineffective if the impure metal contains magnesium which may have come from Al–Mg scrap; due to magnesium reacting with the electrolyte.

8.4 Properties

Atomic weight of aluminium is 26.97, melting point is 660°C and boiling point is 1800°C. The main impurities of commercial aluminium (99.3–99.6% Al) are iron and silicon together with traces of titanium and copper. At room temperature, aluminium has a specific gravity of 2.7 as compared with 8.89 for copper and 7.8 for iron. Its high strength to weight ratio is a distinct advantage in many structural applications. The metal exhibits good corrosion resistance, electrical and thermal conductivity. The thermal conductivities and electrical resistivities of Ag, Cu, Au, and Al (as shown in Table 7.1) are outstanding when compared with those of other metals. The metal has high reflectivity of light and low emissivity. Aluminium has strong affinity for oxygen, a film of oxide forms spontaneously on any freshly cut surface. To this property aluminium owes its high resistance to chemical and atmospheric attack. Aluminium can be hot rolled, extruded, forged, cold rolled, drawn, stamped and machined as readily. One of the primary characteristics of aluminium is that it can be worked and formed by the usual fabricating processes.

The tensile strength of commercial aluminium is low, ranging from 6.3–14.2 kg/mm^2 according to the amount of cold working. The modulus of elasticity of aluminium (and its alloys) is approximately one-third of that of steel, which means that the elastic deflection under load of aluminium is three times that of steel.

Purer grades of aluminium are known as super-purity aluminium, and contain not less than 99.99% Al. The super-pure aluminium is softer, more ductile, heats more efficiently, and possesses higher thermal and electrical conductivities and greater resistance to corrosion than the commercial metal.

The addition of other metals to aluminium affords the means of increasing the strength and hardness. The small amount of impurities present in commercial alluminium is sufficient to increase its strength up to 50%, as compared with the pure metal. The metals most commonly used in the production of aluminium alloys are copper, silicon, manganese, magnesium and zinc. These metals may be added singly or in combination to produce a desired result.

Al–Cu (8%) contains 1% Fe, and 1.2% Si; and Al–Cu (Y alloy) contains 4% Cu, 2% Ni, and 1.5% Mg. Y alloy is superior to first alloy due to addition of Ni. Y alloy was developed to withstand stress at elevated temperatures and is largely used as piston material. Duralumin has the composition 3.5–4.5% Cu, 0.5% Mg and 0.5% Mn; together with small quantities of Si and Fe. It is the first heat treatable alloy to be discovered and age hardening is done at ordinary room temperature for 4–5 days to increase in strength and hardness spontaneously.

Al–Si alloy has high fluidity, possess good casting qualities and also has excellent resistance to atmospheric corrosion, making it of value in the production of architectural and ornamental castings. Alloys containing 5% Si are used for these purposes.

Al–Mg alloys have combined properties of high resistance to marine exposure, excellent strength, ductility and good machining qualities. They are rather difficult to cast and require special treatment to prevent oxidation while casting. Al–Mg (10%) alloy has the highest combination of tensile strength, elongation and resistance to impact, after heat treatment.

8.5 Applications

After copper, aluminium metal is considered the most important non-ferrous metal. Since aluminium is a good conductor of heat and electricity, and it is also less expensive than copper; nowadays it is replacing the copper in electrical applications. In India, aluminium is gradually replacing the copper in the power transmission lines and in other electrical applications. A composite conductor made of aluminium wires that surround a core of steel wires with a high tensile strength is being extensively used as an overhead conductor. The core compensates for the low strength of aluminium. For transformer winding, aluminium strips or foils interleaved with insulators are used. Due to very good ductility and flow properties of aluminium, it is used for extrusion of vessels and containers. For kitchenware, chemical and food processing industries, pure grade aluminium is used due to its corrosion resistance and non-toxic properties. High reflectivity and low emissivity of aluminium explains its use as roofing and thermal insulation either as metal foil or aluminium paint. In steelmaking, aluminium acts as deoxidizing agent to remove oxygen in liquid steel.

Chapter 9
Zinc

9.1 Introduction

Zinc (Zn) has been known from ancient times as an alloy with copper. It was used in India and China in the 5th century. The first real knowledge of smelting comes from India and China, the latter producing zinc on a moderate scale in the early 1700s, a sample of Chinese zinc dated 1745 having been found to contain 98.9% zinc. Production of zinc started in Great Britain in 1740 by the process of distillation. In Belgium the horizontal-retort process was developed in 1806. By-products include cadmium, sulphuric acid and fertilizers. The original Belgian thermal process still remains the basis for modern zinc distillation. In this method the roasted or sintered concentrate (or both) is reduced by strongly heating with coal in vertical retorts. Metallic zinc distills and is collected as liquid metal in condensers. Produced zinc contains 1–2% lead and was formerly known as spelter. By fractionating spelter, 99.99% zinc can be obtained.

9.2 Sources

The important ores of zinc are sphalerite (ZnS) and marmatite [(ZnFe)S]. Less common ores are zincite (ZnO), smithsonite ($ZnO.CO_2$), and franklinite. The oxidized zinc ore calamine ($Zn_2O.SiO_3$) was at one time the main source of zinc. Zinc ore is widely distributed, the most important sources being USA, Canada, Australia, Mexico, Germany, Poland and India. However, with the successful separation of sulphide minerals by flotation, its importance has considerably decreased. Sphalerite (ZnS) and marmatite [(ZnFe)S] are the main economic ores, which are always found with the lead mineral galena; they are separated by flotation.

S.K. Dutta and D.R. Lodhari, *Extraction of Nuclear and Non-ferrous Metals*,
Topics in Mining, Metallurgy and Materials Engineering,
https://doi.org/10.1007/978-981-10-5172-2_9

Zinc ores/minerals:

(1) Zinc blende or sphalerite (ZnS).
(2) Smithstone ($ZnCO_3$).
(3) Calamine ($Zn_2SiO_4.H_2O$).

Zinc ores are often a low grade ores which contain 2–7.5% Zn. By mineral beneficiation process (i.e. flotation), separation of sulphides of Pb, Cu take places and concentration of zinc (ZnS) is produced. Zinc concentration contains 47 to 60% Zn, 1.5 to 2.5% Pb, 3.5% Cu, 3 to 10% Fe, 0.2% Cd and 29 to 33% S; apart from those minor quantities of Tl, In, Ga, Ge, Se and Te are also present.

9.3 Extraction of Zinc

There are two routes of extraction of zinc: pyrometallurgical process, and hydrometallurgical process followed by electrometallurgical process.

9.3.1 Pyrometallurgical Process

The techniques of smelting and converting are applied for Cu_2S in copper metallurgy. But for ZnS, they are not possible due to high melting point of ZnS even at 1500 °C. Hence, a different method is applied in zinc extraction.

Flow sheet for pyrometallurgical process is shown in Fig. 9.1. After grinding of ore, separation of sulphides of lead and copper is done; and concentration of zinc (50–60% Zn) is produced.

9.3.1.1 Roasting

Oxidation of zinc concentrate (ZnS) is necessary for the subsequent reduction and recovery of zinc metal. Oxidation is carried out by roasting, either a partial roast followed by sintering or in one operation as a dead roast. Complete elimination of sulphur is desirable as any remaining after roasting ties up twice its weight of zinc, which is lost in subsequent operation. The zinc sulphide concentrate always contains some lead sulphide which is eliminated during initial stages of roasting, i.e. by volatilization; hence, the multiple hearth roaster is used. The roaster has eight hearths and a top drier hearth. The remaining lead sulphide is removed during sintering at high temperature. Heating is unnecessary when a partial roast is done, for the blende contains sufficient sulphur to support combustion down to 7–8% sulphur; but for dead roasting, external heating is necessary to drive off the last few per cent of sulphur; this is provided by a firebox on the bottom hearth, either oil or gas is used as fuel.

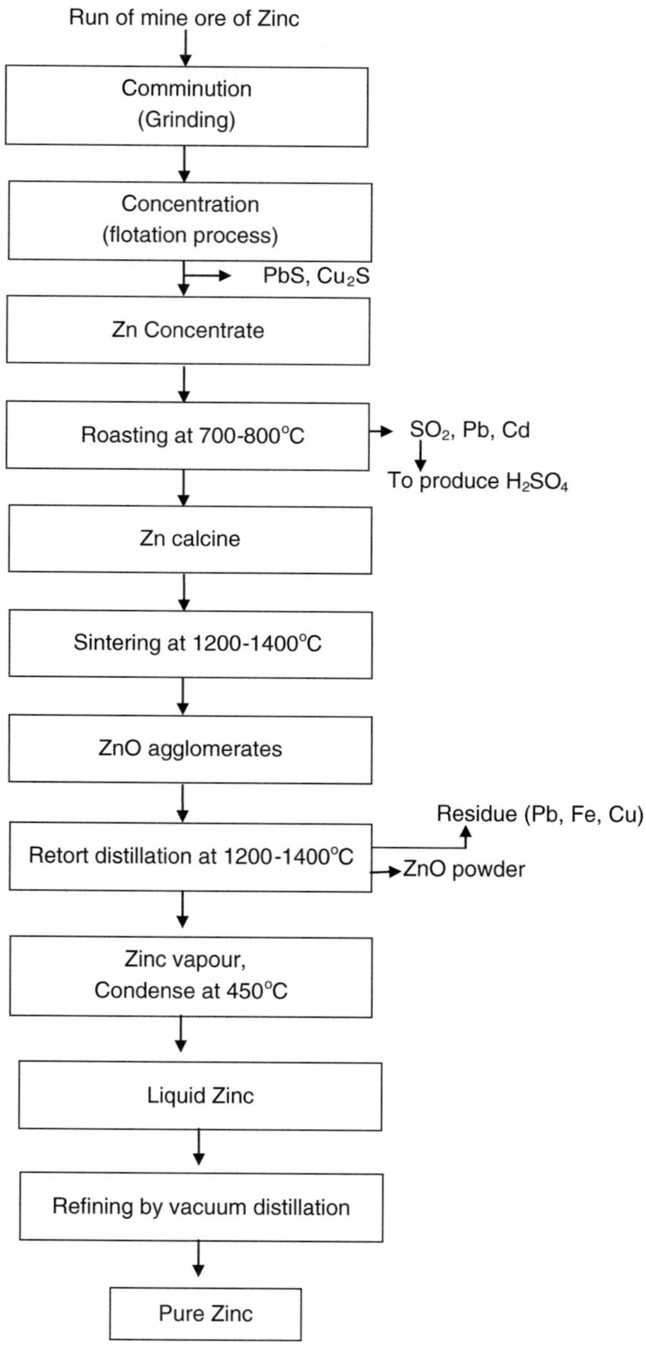

Fig. 9.1 Flow diagram for extraction of zinc by pyrometallurgical process

Partial roasting of zinc concentration (ZnS) is done at 700–800 °C to zinc calcine (ZnO, which contain 1–4% S). For a dead roasting a higher temperature (1000 °C) is required.

$$2ZnS + 3O_2 = 2ZnO + 2SO_2 \quad (9.1)$$

Presently, roasting is carried out in a suspension roaster or fluidized bed roaster instead of earlier in a conventional hearth roaster, because the former has certain advantages: (a) it is autogenous; (b) good control over the temperature; and (c) high roasting rate.

Then sintering of zinc calcine (ZnO) is done at 1200–1400 °C in a Dwight-Lloyd sintering machine to provide lump feed for retort distillation and to eliminate residual sulphur, cadmium and lead.

9.3.1.2 Retort Distillation

In horizontal retort process, 2 m long retorts, each having a diameter of 25 cm and a thickness of 5–6 cm, are lined up in a gas fired furnace in batches of several numbers at a time. Retorts cannot make of metals because most of metals produce alloy with zinc when heated at high temperature. The retorts are generally made of fireclay. The retorts are first charged with mixture of fine size ZnO and an excess of coal (30–40% by weight), then heated to a maximum temperature to volatilize zinc which is subsequently condensed. Theoretically, one part of carbon will reduce seven parts of zinc oxide, but in practice it has been found necessary to use four to five times this amount, principally to keep the carbon dioxide concentration as low as possible ($C + CO_2 = 2CO$), and also to protect the retorts by holding up the molten slag. This large excess of carbon, over that the theoretically required, is necessary to insure complete reduction and to maintain a reducing atmosphere in the retort by reducing any carbon dioxide that may be formed.

The furnace is gradually heated up to 1400 °C, distillation start around 1100 °C. The first gas to come off is coal gas, burning with a luminous flame; as the reduction proceeds, and carbon monoxide is given off, the luminosity diminishes and the flame becomes purple. The beginning of zinc distillation is marked by the flame of bluish-white colour and generally at this stage the prolongs are fitted to the condensers. Finally, when all the distillable zinc has been expelled, the flame loses its bluish tinge and becomes purple again, indicating that the cycle has been completed. Temperature of condenser is important: if it becomes too low (below 400 °C) excessive blue powder forms, and if too high, zinc vapour will pass off uncondensed. A temperature of 1400 °C is generally maintained in the furnace, and retorts temperature being 100–200 °C lower.

Reduction of zinc oxide by carbon is a strongly endothermic reaction, which requires temperature in the range of 1200–1400 °C.

$$ZnO(s) + C(s) = Zn(g) + CO(g)\ \Delta H^{\circ}_{f} = 238.49\,kJ \tag{9.2}$$

The liberated zinc vapour is extremely sensitive to oxidation and it is necessary to carry out the distillation in a closed system. The carbon monoxide forms in reaction (9.2), then further reduces to zinc oxide:

$$ZnO(s) + CO(g) = Zn(g) + CO_2(g)\ \Delta H^{\circ}_{f} = 75.31\,kJ \tag{9.3}$$

At high temperature the carbon dioxide is unstable, so it reacts with more carbon present in the system to form carbon monoxide:

$$C(s) + CO_2(g) = 2CO \tag{9.4}$$

Zinc boils at 907 °C, and carbon does not reduce zinc oxide until 1120 °C, so zinc is liberated in the vapour form. In this state it is not only susceptible to oxidation by the ordinary oxidizing agents such as air and water vapour, but it is capable of oxidizing by carbon dioxide. Reactions (9.3 and 9.4) are reversible reactions, but the concentration must be very high before the backward reactions take place. With zinc vapour [reaction (9.3)], however, the backward reaction occurs with small concentrations of carbon dioxide. In general, condition of reduction is $(p_{co}/p_{co2})_{gas\ phase} > (p_{co}/p_{co2})_{equilibrium}$, so p_{co} should be maintained at a high level in order to prevent the oxidation of zinc vapour to zinc oxide due to presence of CO_2 gas. If the zinc vapour is much diluted with other gases, it condenses not to liquid zinc but to a mixture of solid zinc encrusted with zinc oxide, which is known as blue powder.

9.3.1.3 Vertical Retorts

Vertical retorts are made of carborundum, as refractory by mixing with 6–12% of clay. Carborundum is nothing but silicon carbide (SiC), which has a melting point above 2700 °C. The conductivity of silicon carbide is about five times higher than that of fireclay. Vertical retort is a shaft of rectangular cross-section. The height of the heating zone is about 7.5 m, and inside dimensions of retort cross-section average 300 × 1500 mm.

Generally, gas is used to heat the retort by burning of gases in chambers. Modern vertical retort is heated by electrical energy. The graphite electrodes are introduced through the silicon carbide walls near the bottom of the retort. The electrical energy provides the heat to the system. The charge is introduced as briquettes from the top and slag is discharged continuously from the bottom. Modern vertical retorts are provided with automatic charging and discharging systems which help the handling of raw materials as well as products.

The roasted zinc concentrate, coking coal and coal tar (as binder), after a through mixing, the charge is passed through a briquetting machine. The briquettes are

Table 9.1 Comparisons between horizontal and vertical retorts

Horizontal retort	Vertical retort
1. Fine-sized roasted zinc concentrate and coal or coke act as charge materials	1. Sinter/briquette of roasted zinc concentrate and coal or coke act as charge materials
2. It is batch process	2. It is continuous process
3. It is difficult to mechanize and hence involves large manual labour (29–43 man-hours)	3. It can be easily mechanized and hence involves less manual labour (7.7–12.5 man-hours)
4. It has less capacity (800–1000 kg/day) and less recovery of zinc (86–88%)	4. It has more capacity (7–10 tonne/day) and high recovery of zinc (90–94%)
5. Size of retort is smaller (2 m long, 0.25 m diameter and 0.05–0.06 m thickness)	5. Size of retort is larger (10 m long, and 0.3 x 1.5 m cross-section)
6. Retorts are made of fireclay	6. Retorts are made of silicon carbide

heated to 750–900 °C to strengthen briquettes and make them adequately porous. Distillation in vertical retorts is no different in chemistry than the standard horizontal retort technique. Since the conductivity of silicon carbide is higher and therefore zinc is removed more completely, zinc content in slag is generally less than 5%.

A condenser is lined with refractory brick and provided with internal partitions that lengthen the path of gases. The zinc vapour is condensed very fast in order to avoid the formation of blue powder (ZnO + Zn). 3–4% zinc, all iron, gangue, copper, and precious metals are contained in the residues of the retort processes. Zinc recovery is about 95%. The comparisons between horizontal and vertical retorts are shown in Table 9.1.

Zinc vapours are condensed to metal and some are oxidized. Besides zinc, retort slag always carries lead, copper and noble metals; therefore they require additional treatment. A top layer of zinc containing some lead (about 0.8%) and an intermediate layer of zinc and iron (15–20%) is called hard metal, which is recycled to the retorts. Bottom lead layer is further smelted to recover the lead. The zinc layer is refined by fractional distillation due to wide difference in the boiling points of zinc (907 °C), cadmium (780 °C) and lead (1620 °C); therefore, easy separation of one metal from the other.

9.3.1.4 Physical Chemistry of Zinc Smelting

The reaction taking place in a retort may be written as (in general):

$$(x + 2y)ZnO(s) + (x + y)C(s) = (x + 2y)Zn(g) + xCO(g) + yCO_2(g) \quad (9.5)$$

When zinc oxide and carbon are heated together, variable amounts of zinc, carbon monoxide and carbon dioxide are produced. This reaction (9.5) is the sum of the following reactions:

$$xZnO(s) + xC(s) = xZn(g) + xCO(g) \quad (9.6)$$

$$2yZnO(s) + yC(s) = 2yZn(g) + yCO_2(g) \quad (9.7)$$

It is significant that the so-called temperature of reduction of zinc oxide had been reported by investigators over the range of 780 to 1310 °C. These divergences had been attributed in the past to differences in the physical form of zinc oxide and carbon, to differences in the gas pressure, and to allotropic modifications. It is certain that no amount of continued heating at a definite pressure will cause gas-free carbon to reduce zinc oxide continuously below a certain temperature. It is also obvious that attempts to measure the temperature at which reduction takes place have measured only the temperature at which the rate of reaction becomes measurable. At higher temperatures, where the reaction takes place at any definite temperature, either zinc oxide or carbon, or both, disappear from the system. In other words, if a system containing zinc oxide and carbon is vented at a definite pressure, a condition of equilibrium can exist at only one temperature. Below this temperature, reduction by carbon does not take place, and above it continuous reaction takes place. This is the system which is unstable as far as chemical equilibrium is concerned.

Reactions within the charge in a retort produce water vapour, hydrocarbons, carbon monoxide and carbon dioxide; but only the last two (CO and CO_2) are the important factors in the production of *blue powder*, which is a mixture of zinc and zinc oxide.

Zinc is a moderately strong basic metal and may be used to reduce carbon monoxide:

$$Zn(g) + CO(g) = ZnO(s) + C(s) \quad (9.8)$$

This would necessarily involve the reversal of the producer gas reaction:

$$2CO(g) = CO_2(g) + C(s) \quad (9.9)$$

$$CO_2(g) + Zn(g) = ZnO(s) + CO(g) \quad (9.10)$$

The reaction (9.8) is the combination of the reactions (9.10) and (9.9).

It has been shown that reaction (9.9) is slow below 1100 °C but reaction (9.10) is perceptible above 300 °C and very fast above 550 °C, so that the rate at which zinc is oxidized is determined by the slowest step. Large amounts of carbon dioxide would be present in the equilibrium mixture at temperature below 850 °C, but at this temperature the rate of the reaction is so slow that equilibrium would be obtained only in week or month. Catalysts are known that would increase this rate, but they are not present in the charge. Therefore, the reaction with carbon monoxide probably accounts for the formation of only a part of the blue powder.

9.3.2 Hydrometallurgical Process

(a) Hydrometallurgical processes are used for the extraction of zinc from the low grade ores. These processes produce high purity zinc and also lead to a higher rate of recovery of valuable by-products. Figure 9.2 shows the flow sheet of hydrometallurgical process.

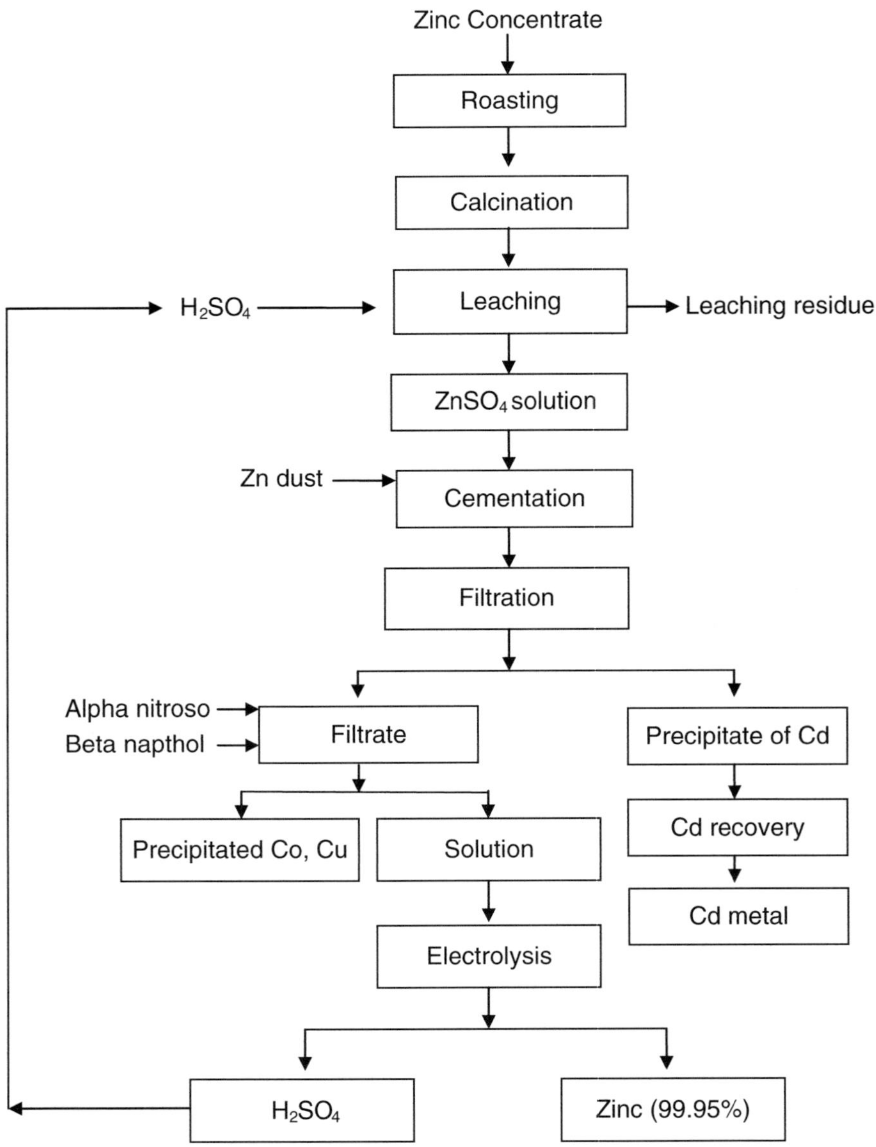

Fig. 9.2 Flow diagram for extraction of zinc by hydrometallurgical process

Before leaching, the zinc sulphide concentrate is first roasted. Roasting for leaching must be very carefully controlled to produce the right proportions of zinc oxide and zinc sulphate. The zinc sulphate must be present in sufficient amount to make up for the sulphuric acid lost in dissolving the oxide and processing the residues. The roasted zinc concentrate is dissolved in sulphuric acid and the valuable by-products are precipitated out.

$$ZnO + H_2SO_4 = ZnSO_4 + H_2O \tag{9.11}$$

Zinc dust is added to the leach zinc sulphate solution to precipitate cadmium (i.e. cementation process) for recovery of Cd; Co and Cu are also precipitated out. Precipitation of cobalt from leach solution is done by adding α-nitroso-β-napthol. Then electrolysis of the purified leach solution ($ZnSO_4$ + H_2SO_4) of 5.3 pH is done by using a cell with an anode of cast Pb-Ag (1%) alloy and rolled high-purity aluminium sheet as cathode. Metallic zinc is deposited at cathode and oxygen gas is released at anode. Cu and Fe present in solution reduce the current efficiency of the electrolyte bath.

(b) Acid-pressure leaching of zinc concentrates was developed in Canada. To obtain high zinc recovery in the leaching process, zinc concentrate should be very fine sizes. Continuous-pressure leaching is carried out in two to three compartments, along with autoclaves. After leaching, the slurry is cooled by two-stage flashing at atmospheric pressure. After a solid–liquid separation and residue is washed properly, the liquor is passed for sulphur removal by neutralization. After neutralization, iron is removed from the leach liquor. In the first stage, the liquor is neutralized at 80–90 °C and pH is maintained at 3.5–4.0 with limestone. Air is spurged into the neutralization reactor to oxidize iron for hydrolyses and precipitation. In the second stage, zinc dross is used for further neutralization to pH 4.5. In the second stage, copper losses are significant; and so that is recycled. A purification process is carried out by zinc dust before coming to electro-winning.

(c) In the conventional process, certain impurities (like selenium, mercury, fluoride, and chloride) are removed during the roasting of the zinc concentrates. In the direct leaching process, selenium, mercury, and fluoride, are removed either in the leach residues or during the iron precipitation steps. It is well known that selenium is difficult to remove from leach liquors of copper concentrates. Since selenium is very harmful in the zinc electro-winning step, separate purification procedures are followed to keep its level low, which is acceptable in the electro-winning cells. Any chloride present in the zinc leach liquor has to be removed prior to electro-winning by either ion exchange or silver chloride precipitation.

9.4 Properties

Atomic weight of zinc is 65.38, specific gravity is 7.14; melting point and boiling point are 419 °C and 907 °C respectively. The relatively high corrosion resistance of zinc maks it useful as a protective coating (galvanizing) on steel. Zinc, being electro-negative to iron, is slowly corroding zinc by corrosive media, leaving the iron undamaged and protects iron. If the coating is damaged at any point, its protective properties remain unaffected, since base sections of iron are not attacked. On cast condition, zinc possesses a tensile strength of 3.15 to 7.87 kg/mm^2; after cold work that is increased to 18.9–25.2 kg/mm^2.

Zinc dissolves in strong acids (like HCl, H_2SO_4) with an evolution of hydrogen. Zinc oxide (ZnO) is a fine crystalline white powder obtained by oxidation of zinc vapour in the air. Zinc oxide is a non-fusible substance which volatilizes at a temperature above 1800 °C without melting. Reduction of zinc oxide can be done by carbon at the temperature 950 °C.

9.5 Applications

After cold work, tensile strength is drastically increased; hence, zinc is not used where high stresses are likely to be encountered; but it can be used for ordinary applications such as roofing, metal linings, dry batteries etc. By rolling, the metal is easily worked to make foil, due to ductility of zinc.

High purity zinc (99.99%) is used for manufacturing of die casting parts, wire, foil and anodes. High grade zinc (99.95%) is used for production of wire, zinc oxide, and high grade brasses. Foundry grade zinc (98.5%) is used for making sheet and strip by rolling; foundry work for brass and other alloys castings. Zinc (98%) is used for galvanizing, formation of zinc dust and brass. Zinc acts as an alloying element in brass and bronze. The bulk of the metal is consumed as an alloying element of brass and in the form of galvanized steel sheet.

Small articles are exposed to zinc dust at elevated temperatures (350–400 °C); zinc diffuses at the surface, and this method is known as sherardizing. This process is used for steel windows. Springs, aeroplane components, bolts, nuts and small castings of intricate design can be made from zinc and its alloys. Almost any metal may be applied by spraying, since zinc has a low melting point (420 °C) and corrosion prevention property, hence being used to a greater extent than any other metal. Anti-corrosion paints based on metallic zinc powder form, a convenient alternative to metal spraying.

On account of low melting point and prevention of corrosion, zinc is used to spray on the other metals. Zinc in the form of wire or powder is loaded into a pistol, where it is melted and atomized by compressed air; then is projected at the target object. The main applications of sprayed zinc coating are for steel bridges, ships, storage tanks etc.

The copper-zinc alloys can be divided into two main groups: (a) first group contains 30–37% Zn, which has adequate ductility and strength for cold deformation into sheets and wire; (b) second group contains 37–45% Zn, which is much less plastic when cold and is therefore worked above 500 °C. The zinc die-cast alloys contain small amounts of Al, Cu and Mg. The most widely used of these zinc die-casting alloys contain up to 4% Al with small amounts (0.02%) of Mg.

Zinc oxide is used for high grade pigments and pharmaceutical use. Zinc oxide is extensively used in the manufacture of oil and other types of paints owing to its white colour and high covering power.

Chapter 10
Lead

10.1 Introduction

Lead (Pb) has been known since ancient times, as its use from the 7th to the 5th millennia B.C. is now a well established fact. Since the dawn of civilization, lead has served for the conveyance of water. The bath erected by the Romans in the city of Bath was lined with sheet lead and it is claimed that it is still supplied with water through the original lead pipes which were installed nearly 2000 years ago. In later years the use of lead was extended to include roofing, as witness many churches and cathedrals of the Norman period. Since that time lead has been called upon to serve many other purposes where corrosion resistance is needed.

Reverberatory furnaces for lead smelting were used until late in the 1800s. The process consisted of a roast-reduction, oxide and sulphide formed by roasting the ore was reduced by unaltered sulphide. Slag contained 40% lead and this was dealt with in the slag hearth, a modified, small blast furnace. Such a method was prevalent in the UK, Europe and elsewhere until superseded by the blast furnace, which now produces a majority of the world's lead.

10.2 Sources

Lead is produced from ores, and it occurs in the form of various minerals; galena (PbS) is the most common ore of lead and zinc. Other minerals are anglesite ($PbSO_4$) and cerussite ($PbCO_3$). Galena is associated with ZnS, FeS, Cu_2S, SiO_2, other impurities (Bi, As, Sb) sulphides and precious metals (mainly Ag). Ores are generally of a complex variety, chief lead associates being zinc, copper and silver. Content of zinc often exceeds that of lead.

S.K. Dutta and D.R. Lodhari, *Extraction of Nuclear and Non-ferrous Metals*,
Topics in Mining, Metallurgy and Materials Engineering,
https://doi.org/10.1007/978-981-10-5172-2_10

10.3 Extraction of Lead

Lead sulphide, at the temperature ordinarily reached in practice, is not reduced by carbon or carbon monoxide; and consequently lead sulphide ore must be roasted for smelting. Furthermore, the presence of too much sulphur in the ore charge would tend to throw an undue proportion of the lead into the matte and render the subsequent recovery of the metal difficult. Since most lead ores contain small amounts of copper, every effort is made to concentrate this in the form of a matte. On the other hand, there should be only a small amount of this matte; hence the necessary of regulating the sulphur content carefully by means of preliminary roasting. In principle, roasting lead ores differs only slightly from roasting of copper ores. It should be noted that galena (PbS) is much more easily fused than most copper minerals, and consequently the temperature must be more carefully regulated.

Flow diagram of extraction of lead is shown in Fig. 10.1. Method of lead extraction from galena involves the following stages:

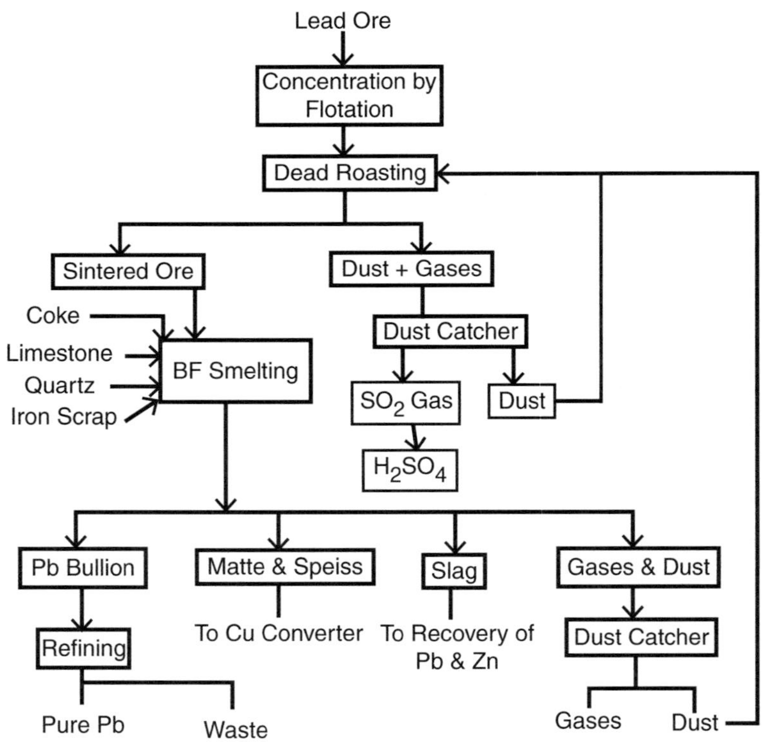

Fig. 10.1 Flow diagram of extraction of lead

(1) Concentration.
(2) Dead roasting.
(3) Smelting.
(4) Refining.

10.3.1 Concentration

The theoretical lead contents of lead sulphide is 86.8%, but actually lead contains much less in ore. Zinc and other sulphides are separated by flotation. Lead ore is concentrated by flotation; lead is enriched up to 60–80%. Silver almost invariably follows the lead into the concentrate, remaining with it during smelting; it is finally recovered during the refining operation.

10.3.2 Dead Roasting

The main object of the roasting of lead concentrate is to eliminate or carefully control the percentage of sulphur in the ore. It is done in a Dweight-Lloyd sintering machine. During roasting, the temperature (at 800°C) must be carefully controlled because, above 800°C, PbS tends to fuse. This machine not only eliminates sulphur to a large extent but also produces a sinter with required properties, such as strength and porosity, for the subsequent blast furnace operation. Sulphur is reduced from 16–18% to 1–2%. The flue gases contain 1–3% SO_2 depending upon the nature of the charge and operating conditions.

Galena ore was first roasted at a temperature below the fusion point in order to promote formation of oxide and sulphate.

$$2PbS + 3O_2 = 2PbO + 2SO_2 \tag{10.1}$$

$$2FeS + 3O_2 = 2FeO + 2SO_2 \tag{10.2}$$

At lower temperatures, sulphates can be formed:

$$PbS + 2O_2 = PbSO_4 \tag{10.3}$$

Fe_2O_3 and SiO_2 are always present in raw materials, hot Fe_2O_3 and SiO_2 act as catalysts to promote the oxidation of the sulphur dioxide to trioxide and that may react with lead oxide to form the lead sulphate.

$$2SO_2 + O_2 = 2SO_3 \tag{10.4}$$

$$PbO + SO_3 = PbSO_4 \tag{10.5}$$

10.3.3 Smelting

A blast furnace (BF) is used for smelting of lead ore sinter. This BF is smaller than one used for smelting copper or nickel. Height of BF is 8 m, top diameter is 3 m and stack height is 5 m. Coke, acting as fuel and reducing agent, is charged (10%) along with sinter from sintering machine; limestone and quartz (10%, act as flux), and iron scrap (1%, to reduce lead losses in the slag) are also charged in BF. Cold blasts of air are blown in through tuyeres (15–20 numbers) around the bosh to burn coke and oxidize the carbon (which is present in coke), producing the required amount of heat. The major portion of the heat is furnished by the combustion of coke, because all the sulphur which is desired to be eliminated has been oxidized in the earlier roasting operation. The bosh is water-cooled to ensure that the bosh temperature and the temperature of the outgoing gases remain low. This is to reduce the loss of lead due to vaporization (boiling point is 1737°C).

Burning of coke at tuyeres region:

$$C + O_2 = CO_2 \ \Delta H_f^\circ = -393.5 \text{ kJ/mol of C} \tag{10.6}$$

$$CO_2 + C = 2CO \ \Delta H_f^\circ = 172.6 \text{ kJ/mol of C} \tag{10.7}$$

Overall reaction: $2C + O_2 = 2CO$

or

$$C + 1/2O_2 = CO \ \Delta H_f^\circ = -110.45 \text{ kJ/mol of C} \tag{10.8}$$

Reduction of lead oxide and other oxides are done by counter current motion of carbon monoxide gas at BF:

$$PbO + CO = Pb + CO_2 \tag{10.9}$$

$$As_2O_3 + 3CO = 2As + 3CO_2 \tag{10.10}$$

$$Sb_2O_3 + 3CO = 2Sb + 3CO_2 \tag{10.11}$$

$$Cu_2O + CO = 2Cu + CO_2 \tag{10.12}$$

The iron oxide is also reduced by CO, and this iron together with scrap iron (which is added as part of the charge) is also reacted with lead sulphide and lead oxide.

$$FeO + CO = Fe + CO_2 \tag{10.13}$$

$$PbS + Fe = Pb + FeS \tag{10.14}$$

$$PbO + Fe = Pb + FeO \tag{10.15}$$

Metallic lead is also formed by the interaction between lead sulphide and oxide or sulphate; as well as by direct reduction by carbon.

$$2PbO + PbS = 3Pb + SO_2 \tag{10.16}$$

$$PbSO_4 + PbS = 2Pb + 2SO_2 \tag{10.17}$$

$$2PbO + C = 2Pb + CO_2 \tag{10.18}$$

As the temperature approaches 900°C, the limestone of the charge is decomposed; liberating lime, which is united with the silica and other gangue materials present in the system, to form a slag.

$$CaCO_3 = CaO + CO_2 \tag{10.19}$$

$$\text{Slag formation: } CaO + SiO_2 = CaO.SiO_2 \tag{10.20}$$

$$2PbO + SiO_2 = 2PbO.SiO_2 \tag{10.21}$$

$$2PbSO_4 + SiO_2 = 2PbO.SiO_2 + 2SO_3 \tag{10.22}$$

$$2FeO + SiO_2 = 2FeO.SiO_2 \tag{10.23}$$

As atmospheric oxygen plays an important role in the reduction, enriching the air blast with 2% oxygen has increased output by about 15%, as well as saving in coke and smoother operation. Another important factor in securing efficiency lies in the production of a suitable slag, which depends upon the rate of smelting and recovery of lead and silver. In order to secure fast running, a satisfactory separation of the metal and slag must take place, and to ensure this, certain requirements of the slag are necessary as follows:

(1) Low specific gravity, for the greater the difference in density between it and the metal or the matte, the more rapid and complete the separation.
(2) Low melting point, for to flow freely from the furnace, slag must be completely molten.
(3) A slag may have a low melting point and low density, yet if viscous in nature, separation of metal and slag will not be clean, the slag carrying away entrained metal.

Blast furnace slag consists essentially of iron and calcium silicates containing: 18–35% SiO_2, 30–38% FeO, 10–20% CaO, 2–4% Pb plus small amounts of alumina, magnesia, sulphur, etc.

Iron scrap in the charge serves as follows:

(a) It helps in reducing any residual lead sulphide left unroasted during the sintering.

$$PbS + Fe = Pb + FeS \tag{10.14}$$

(b) It combines with silica present in the charge to form $2FeO.SiO_2$, which lowers the melting point of the slag, to increase fluidity at operating temperature.

$$2FeO + SiO_2 = 2FeO.SiO_2 \tag{10.23}$$

(c) It lower the lead losses in slag i.e. it recover lead from slag.

$$PbO + Fe = FeO + Pb \tag{10.15}$$

$$2PbO.SiO_2 + 2Fe = 2FeO.SiO_2 + 2Pb \tag{10.24}$$

At the same time, the sulphur is joined with the copper and iron present to form a matte. At the temperature of about 1000°C, the matte and slag are completely liquefied, coming down to the hearth; where they are separated into the layers according to the specific gravities. In certain cases neither matte nor speiss is formed. If the charge content low sulphur any copper present, that goes with the lead, there is no formation of matte. Production of speiss is closely allied with the amount of arsenic and iron present in the blast furnace.

$$As_2O_3 + 8FeO + 11C = 2Fe_4As + 11CO \tag{10.25}$$

If low in arsenic or antimony, no speiss is formed. If there is an unduly high amount of arsenic, a speiss may form, but ordinarily arsenic content is low enough to be absorbed by the matte or metal. Hence, products of blast furnace are lead bullion, matte and speiss, and slag. Due to difference in specific gravities it gives four distinct layers as follows:

(1) 1st layer: Slag (specific gravity of 3.6, contain: FeO, SiO_2, CaO, MO).
(2) 2nd layer: Matte (specific gravity of 5.2) contains sulphides of copper and other elements (Cu_2S. PbS, CaS, FeS).
(3) 3rd layer: Speiss (specific gravity of 6.0) contains compound of arsenic (if arsenic is present in the charge), Fe_4As and impurities.
(4) 4th layer: Lead (specific gravity of 11.0), it is also called *base bullion*.

Lead (or base) bullion contains 92–99% Pb, 0.1–7% Cu, 0.05–0.8% Ag, $<$ 0.05% Au, up to 1.5% Zn, As, Sb; up to 0.3% Bi and Fe. Presence of impurities has adverse effect on properties, e.g. As and Sb make the lead metal very hard. Lead bullion contains greater part of the impurities in mechanical suspension (rather than in alloy form) due to their higher melting points than lead's.

10.3.4 Refining

Flow sheet for refining of lead bullion is shown in Fig. 10.2. The steps of refining are as follows:

(1) Drossing and skimming.
(2) Softening of lead bullion.
(3) Desilverization.

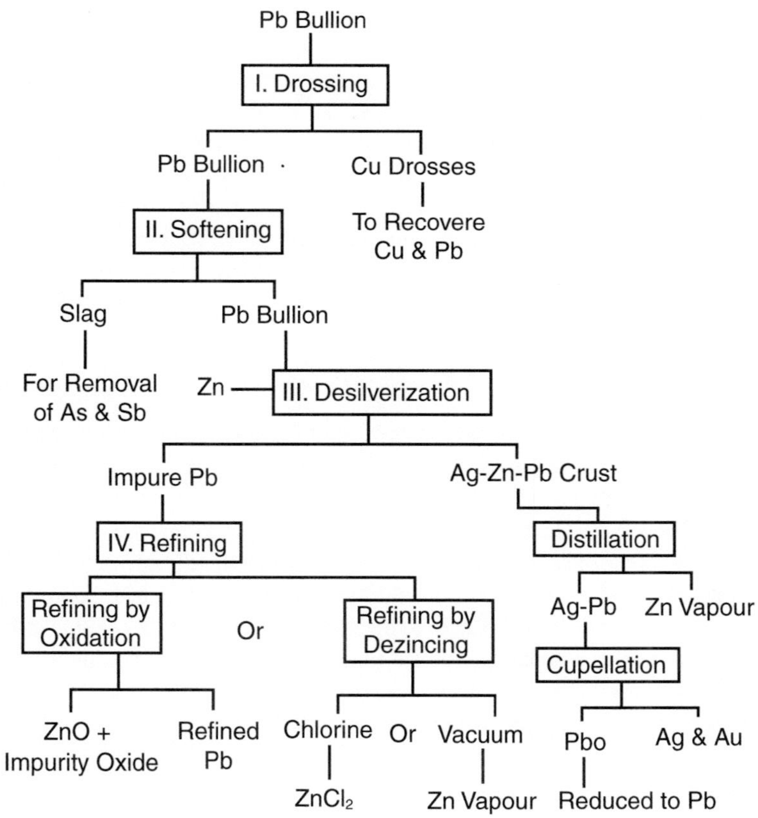

Fig. 10.2 Refining of lead bullion

10.3.4.1 Drossing and Skimming

Drossing is a process of oxidation and liquation. The lead bullion as it leaves the blast furnace well holds the greater part of the impurities in mechanical suspension (rather than in alloy form); and possessing a higher melting point than lead, a lowering of the temperature results in the insoluble oxidized impurities rising to the surface as a dross which can be removed by skimming. The bullion is held at a temperature just above the melting point of the alloy (370–380°C) under the oxidizing atmosphere for a period of about 2 h; during which a dross consisting copper, sulphur, arsenic and lead rises to the surface. This product (copper skimmed) is removed from the furnace for further treatment. Sometimes sulphur is also added into the drossing kettle to eliminate copper as sulphide, copper content is reduced to 0.005%.

10.3.4.2 Softening of Lead Bullion

The drossed bullion, still containing 1–1.5% impurities consisting of arsenic, antimony and tin, is treated in a softening furnace maintaining strong oxidizing condition. The temperature of the bath is raised to about 650°C and air is passed freely to the furnace so that oxidizing conditions may prevail. The lead, arsenic and antimony are oxidized to their respective oxides and form a litharge slag. If tin is present, it also joins this slag; the bath is held at that temperature for about 12 h, during which time nearly all the arsenic, antimony and copper can be removed. Bismuth cannot be removed from the lead by oxidation

10.3.4.3 Desilverization

At the third stage, desilverization is done to lead bullion by addition of zinc. Silver dissolves in all proportions in lead and zinc, with more affinity for the latter. In addition to silver, zinc has affinity for copper and gold, and these two have greater affinity than for silver. Hence the elimination of copper is needed in the drossing and softening operations. Lead and zinc, on the other hand, are only soluble to a very limited extent in each other. Thus, at the melting point of lead, zinc dissolves 0.6%, and lead in zinc to about 1.5%. When zinc is mixed with molten lead bullion and allowed to stand, a separation takes place; the lower layer contains lead with 0.6% Zn; and upper layer of zinc with 1.5% Pb and all silver. Now impure lead, containing 0.6% Zn and a small amount of impurities, is further refining by two methods

(a) Refining by oxidation at 750°C for the removal of zinc. Refined lead, contains 99.98% Pb, is casted.
(b) Refining by dezincing method, there are also two methods: by chlorination to form $ZnCl_2$; by vacuum at 540°C to form zinc vapour.

After third-stage desilverization, the Ag–Zn–Pb crust (consists of 30–60% Pb, 20–50% Zn and 25% Ag) is getting. This is then distilled at 1100°C to separate zinc vapour and Ag–Pb. Cupellation is done in a small furnace, melting the silver containing lead and exposing it to a blast of air whereby the lead and other metals present are oxidized and slagged off. Since silver does not have any affinity for oxygen, it remains on the hearth. By cupellation (oxidation) of Ag–Pb, PbO and silver, gold is separated. Further lead is recovered by reduction of PbO.

10.3.4.4 Electrolytic Refining

In the electrolytic refining process (Fig. 10.3) anodes are made from the lead bullion of blast furnace. Cathodes are thin sheets of pure lead supported on copper bus bars in a cell containing a solution of lead fluosilicate in hydrofluosilicic acid (H_2SiF_6). Lead is dissolved by the electrolyte from the anode and deposited on the cathode, the impurities (such as bismuth etc.) remaining behind (on the anodes).

Finding a suitable electrolyte in which to carry out the electrolysis for refining of lead is a problem. Most of the salts of lead when electrolyzed give rise to side reactions, since the anion, when discharging, is about oxidation, lead peroxide is deposited at the anode. The salt chosen must be readily soluble and the solution must have high conductivity; otherwise excessive electrical energy will be consumed in overcoming the resistance of the electrolyte. Electrolyte consists of a solution of lead fluosilicate in hydrofluosilicic acid. It is formed by the action of hydrofluoric acid on pure silica, resulting in hydrofluosilicic acid which is combined with lead to form lead fluosilicate ($PbSiF_6$).

$$SiO_2 + 6HF = H_2SiF_6 + 2H_2O \tag{10.26}$$

$$H_2SiF_6 + Pb = PbSiF_6 + H_2 \tag{10.27}$$

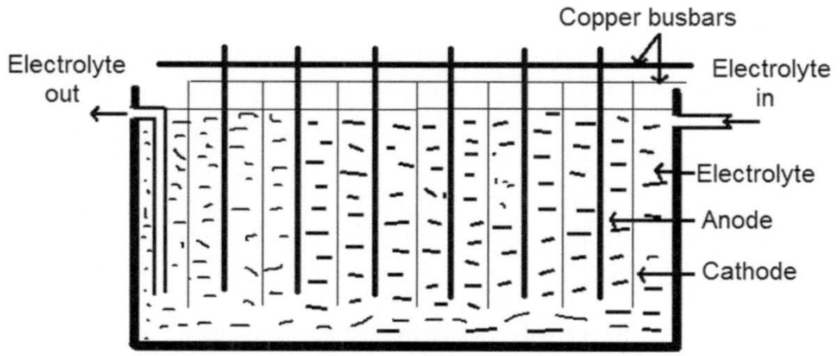

Fig. 10.3 Electrolytic refining of lead

The electrolyte contains 7–10% lead, 10–15% total hydrofluosilicic acid with 5–10% free hydrofluosilicic acid. The composition of the electrolyte necessarily influences the efficiency of the operation, for as the conductivity is dependent upon the presence of hydrogen ions, an increase in the free acid content will increase the number of hydrogen ions with a consequent decrease in the resistance to the current. The presence of fluosilicate increases the resistance, and thus to ensure an electrolyte of minimum resistance it is necessary to have adequate free acid present.

10.4 Properties

The atomic weight of lead is 207.21. Although lead is one of the heaviest of metals (specific gravity: 11.37), it is also one of the softest of the commercial metals. Lead's low melting point (327°C) makes for easy casting. Its boiling point is 1750° C. The low tensile strength (1.575 kg/mm^2) limits its usefulness. For water-carrying purposes, its tensile strength is less important than its creep strength. Metals which are subjected to a steady load stretch slowly in the course of time (especially at high temperatures). This slow stretching or yielding is known as creep. The internal water pressure on pipes exerts a steady load and hence the wall thickness of the pipe must be of sufficient strength to resist continuously any creep tendencies. Lead can easily be rolled for sheets production, and extrusion is used for pipes. The process of joining lead, without use of soldering material, is known as lead burning. The edges of the material to be joined are melted together by means of a small flame, extra lead being added as necessary. Lead burning is invariably used in chemical plumbing. Joins in pipe work, lead to lead, or lead to brass are also carried out. Softness of lead is influenced by impurities.

Corrosion resistance of lead to the action of acids, alkalis and many other compounds promoted its wide use in the protection of chemical apparatus at the time when stainless steels and plastics were still unknown.

In nitric acid, lead dissolves vigorously according to reaction:

$$3Pb + 8HNO_3 = 3Pb(NO_3)_2 + 2NO + 4H_2O \tag{10.28}$$

and at a lesser rate in concentrated acid as compared to a slightly diluted one, because of a poor solubility of $Pb(NO_3)_2$ in a strong acid.

10.5 Applications

In addition to resistance to corrosion by water and weathering, lead also exhibits non-corrosiveness in contact with many mineral acids, hence its extensive use in storage tanks for sulphuric acid. Soft lead (containing 99.99% Pb) is used where lead of high purity is necessary. Antimony and arsenic are often added to the lead to impart hardness for special purposes in chemical plant equipment and lead shots.

The presence of small amounts of other metals in lead modifies and improves certain properties. Addition of 0.003–0.005% each of silver and copper results in an increase in tensile strength of 15%, together with better creep and fatigue resistance. These improved properties are used in pipes for conveying water under pressure; due to these improvements, thickness of the pipe can be reduced. Sn−Cd−Pb alloy is formed of 1.25–1.75% Sn and 0.2–0.3% Cd; it possesses improved creep resistance and tensile strength, but also increased resistance to fatigue under service conditions of vibration. Lead when alloyed with 0.05–0.1% tellurium has the property unique among lead alloys of being amenable to work hardening. By reason of this property, the material strengthens very appreciable when subjected to stress, becoming both tougher and stronger.

The important industrial alloys are those formed by lead with tin, antimony and copper; and, according to their application, these alloys can be classified as follows: bearing metal, antimonial lead, solders.

1. Bearing metal: Various types of bearing metals are available: (a) lead base babbitt; (b) alkali hardened lead bearings; (c) bearing bronzes.

   (a) Lead base babbitt: The term *babbitt* (named after Isaac Babbitt who invented a high tin-bearing alloy in 1839) is generally used to include alloys ranging in composition from 90% or more tin with no lead to 80% lead and less than 5% of tin. The lead base babbitt metals are based on the lead-antimony-tin system and like the tin base babitts have a structure consisting of hard crystals in a relatively soft matrix. The nature of the hard crystals and the matrix depends upon the tin content, and most of the alloys in commercial use fall into one of the two classes. First, those containing 12–18% Sb and up to 5% Sn in which the crystals consist of an antimony-rich solid solution; and second, those containing about the same amount of antimony but with 10–12% Sn. In these alloys the primary phase is SbSn.
   (b) Alkali hardened lead bearings: This class of alloy has been developed as a substitute for high tin alloys rather more for economic than technical reasons. The best known is Garman *Bahnmetall* which contains small amounts of alkali metals (0.6% Na, 0.7% Ca, 0.04% Li) as hardening elements. Another of these alloys is *Satco* containing 0.5–0.75% Ca, 1.5–2.0% Sn, with remainder lead.
   (c) Bearing bronzes: The addition of lead to 10% tin bronze improves the bearing properties and enhances the pressure tightness of the castings. The amount of lead added varies between 5 and 25%, the more heavily leaded bronzes being used where lubrication may be deficient, the lead itself serving in some degree as a lubricant. An important class of copper-lead alloys containing up to 30% lead and only small quantities of tin has been developed for the bearings of diesel and aircraft engines and for other heavy duty purposes.

2. Antimonial lead: The addition of antimony to lead confers the necessary strength and hardness needed in many industrial applications such as chemical plant equipment, battery plates etc. The hard lead, comprising 8–12% Sb, may be machined and screw threaded.
3. Solders: The use of tin-lead alloys to join metals was known some 2000 years ago. Tin and lead are miscible in all proportions and practically the whole range of soldering alloys from 5% to 95% Sn finds use in various industries. The characteristics of solders depend not so much on the mechanical properties as on such physical qualities as wetting ability and penetrative power. Without wetting of a metal surface by a solder there can be no soldering action but only a mechanical anchorage. The molten solder wets or adheres when it leaves a continuous permanent film on metal surface. Wetting is greatly assisted by the ability to alloy as shown by the contrasting behaviour of lead in the manner of solder. Lead when applied to iron or copper fails to wet them but effects joining merely by penetrating and solidifying.

Chapter 11
Tin

11.1 Introduction

The middle of the nineteenth century saw Cornwall in England as the centre of the world's tin industry. Before the Second World War, tin smelting was entirely in the hands of the British and Dutch, but when the Japanese overran the Far Eastern deposits, the USA came into the picture by constructing a large tin recovery plant to treat the available ores from Bolivia in South America.

Cans made of steel that are used for storing edibles are invariably coated with tin. This is due to tin has good resistance properties of corrosion and toxicity. This particular application is universal that a can is often known as *tin*. Apart from plating, tin finds many other applications like as alloying element in bronze, white metal etc., and as soldering element.

11.2 Sources

An important mineral of tin is cassiterite (SnO_2). It is found mainly in Bolivia, Malaysia and Indonesia. Cassiterite is usually associated with gangue minerals and metallic sulphides (PbS, ZnS, $CuFeS_2$ and FeS_2). Theoretical Sn content in cassiterite (SnO_2) is 78.6%, but practically its content is about 1% Sn or even less.

11.3 Extraction of Tin

The following steps are involved for extraction of Sn from tin ore (Fig. 11.1):

(1) Concentration.
(2) Reduction of concentrate.
(3) Treatments of slags for recovery of metals.
(4) Refining of impure metal.

S.K. Dutta and D.R. Lodhari, *Extraction of Nuclear and Non-ferrous Metals*,
Topics in Mining, Metallurgy and Materials Engineering,
https://doi.org/10.1007/978-981-10-5172-2_11

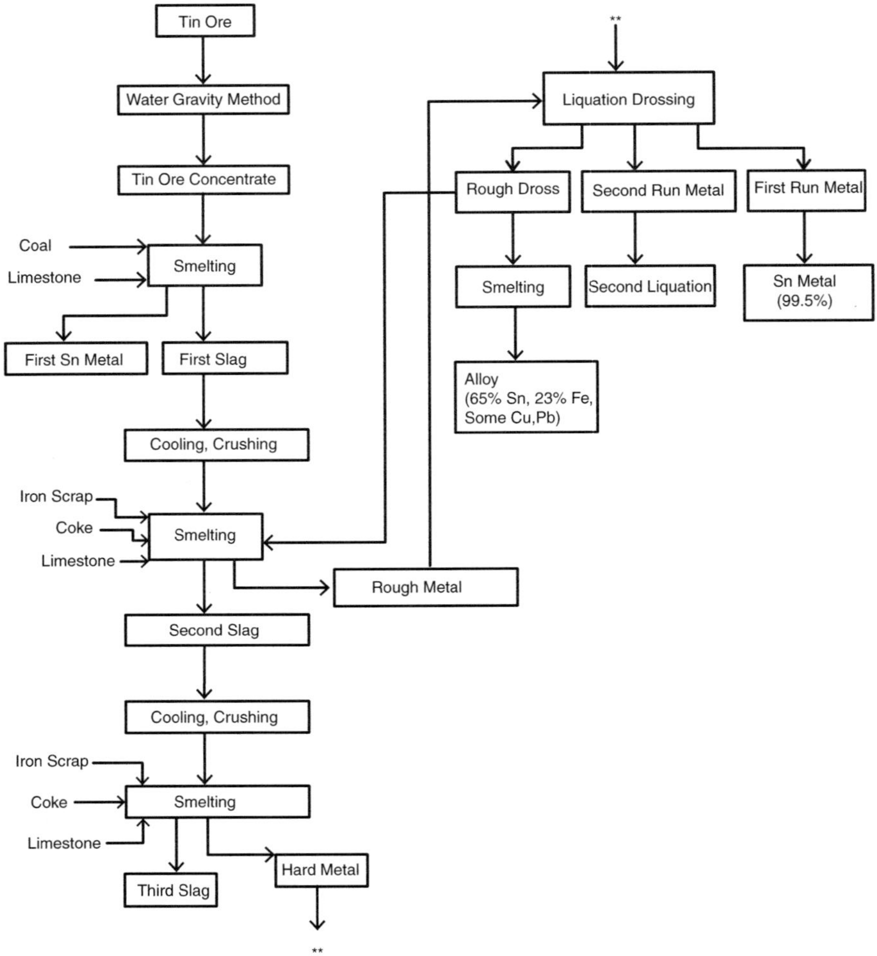

Fig. 11.1 Flow diagram of extraction of tin

11.3.1 Concentration

Tin ore, the content of which is about 1% Sn, is concentrated up to 65% Sn by the water gravity concentration process. At 600°C, SnO_2 and Fe_3O_4 are stable (from Ellingham diagram, Fig. 11.2), at higher temperatures (1200–1300°C) FeO is stable. The slag has a higher affinity for SnO_2 compound with iron oxide and because of the tendency of Fe for entering the tin phase. To overcome these difficulties and to attain a high degree of separation of tin from iron, leading to a high level of tin recovery, smelting of tin concentrate is done in three stages (reduction of concentrate and treatments of slags).

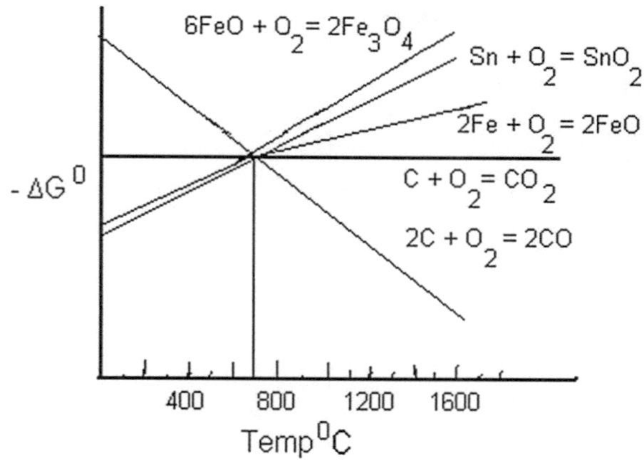

Fig. 11.2 Ellingham diagram for SnO_2, FeO, Fe_3O_4 etc

11.3.2 Reduction of Concentrate

Then smelting reduction of concentrate is done by addition of coal (as reducer) and limestone (as flux) at 1300°C for 10 h. High purity metal (99% Sn) is produce initially and deliberately maintaining a high level of SnO_2 in the slag (which contains 30% Sn), and all iron goes to slag (35% SiO_2, 30% CaO, 15% FeO, 20% SnO_2).

$$SnO_2 + 2C = Sn + 2CO \tag{11.1}$$

11.3.3 Treatments of Slags for Recovery of Metals

This first slag is cooled, crushed and then again smelting reduction is done by addition of iron scrap, coke, limestone and rough dross at 1300°C for 16 h. Rough metal (95% Sn, 5% Fe) and second slag (4% Sn) are formed. The second slag is again cooled, crushed and further smelted at 1300°C with the addition of iron scrap, coke, and limestone. Hard metal (80% Sn, 20% Fe) and third slag (1–2% Sn) are produced.

Sn–Fe alloy produced in the second and third smelting stages can be refined by liquation. Alloy is heated at higher temperature than melting point of tin (232°C). Tin (first run metal) flows out leaving behind Fe–Sn intermetallic. As temperature increases iron content in tin metal is also increased (second run metal).

Liquation drossing of hard metal (80% Sn) is done with the addition of rough metal (95% Sn). Rough dross, first run metal and second run metal are produced by liquation drossing. By smelting of rough dross, alloy (65% Sn, 25% Fe and some Cu, Pb) is again produced. Tin (> 99.5%) is acquired from first run metal. Second run metal again goes for second liquation.

11.3.4 Refining

Flow diagram of tin refining is shown in Fig. 11.3.

(a) **Fe removal**: Intermetallic compound is formed from common impurities, such as: Ca_3Sn (melting point 675°C), Sn_3As_2 (596°C), FeAs (1030°C), Fe_2As (919°C), Cu_3As (825°C), Cu_2Sb (585°C), $FeSb_2$ (726°C), $FeSN_2$ (very high temperature). Cooled to a temperature above melting point of tin (232°C), intermetallic compound is separated out and floats as tiny crystal. Coal helps to float as well as maintain reducing atmosphere. At 550°C, 99.8% Fe is removed from tin metal. Dry dross (content 25% Sn) is formed.

(b) **Cu removal**: Copper is removed by addition of sulphur ($2Cu + S = Cu_2S$), remaining iron is also removed in this stage.

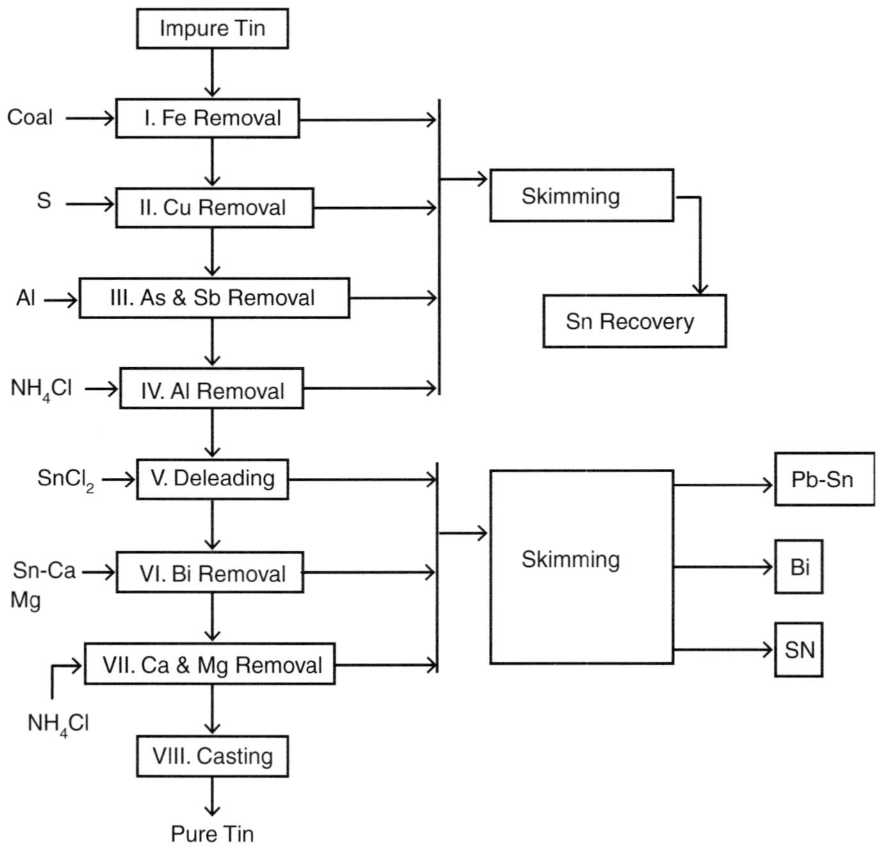

Fig. 11.3 Flow diagram of tin refining

(c) **As and Sb removal**: Arsenic and antimony are removed by addition of alluminium. Alluminium forms intermetallic with arsenic and antimony, such as AlAs (1720°C) and AlSb (1070°C).

(d) **Al removal**: Alluminium is removed by addition of NH_4Cl:

$$Al + 3NH_4Cl = AlCl_3 + 3NH_3 + 3/2H_2. \qquad (11.2)$$

(e) **Pb removal**: Lead is removed by addition of $SnCl_2$:

$$Pb + SnCl_2 = PbCl_2 + Sn. \qquad (11.3)$$

Dross forms [$PbCl_2$ + $SnCl_2$ (excess)], which is reduced by zinc to form Pb–Sn alloy:

$$PbCl_2 + SnCl_2 + 2Zn = 2ZnCl_2 + Pb - Sn. \qquad (11.4)$$

(f) **Bi removal**: Bismuth is removed by addition of calcium and magnesium to form intermetallic compound (Bi_2Mg_5, Bi_2Ca_3 and Bi_3Ca), from them Bi can be recovered.

(g) **Ca and Mg removal**: Excess calcium and magnesium are removed by addition of NH_4Cl:

$$2NH_4Cl + Ca\,(Mg) = Ca(Mg)Cl_2 + 2NH_3 + H_2. \qquad (11.5)$$

(h) **Casting**: Casting of refined tin metal is done as per requirement.

11.4 Properties

The atomic weight of tin is 118.7 and specific gravity of tin is 7.29. Tin has a very low melting point (232°C), but a very high boiling point (2270°C). It is toxic in nature. It can resist corrosion. Cans are made of steels with coated tin on them to prevent corrosion of steels. As tin is also resistant to attack by the common organic acids, it is a safe and attractive medium for use in contact with foods. In point of fact only about 1.4% of the weight of a can consists of tin as coating.

Pure tin is a soft metal with high malleability and low tensile strength; small amounts of impurities, however, tend to modify these characteristics. Iron renders the metal hard; copper makes it both harder and stronger. The softness and plasticity of pure tin make it adaptable to all types of cold working, where high tensile strengths are not required. It may be rolled with heavy reduction without the necessity for annealing, and rolling can be continued down to the thinnest foil.

In many of its properties tin resembles lead, hence the use of the cheaper metal in such materials as roofing, sheets and collapsible tubes for toilet preparations. Tin exists in two allotropic forms, the normal white lustrous metal and a grey brittle modification. When exposed for a sufficient time to low temperatures the white modification changes to the grey powdery form.

Tin in the cast form is strongly crystalline, which gives rise to another peculiarity. When a bar of the metal is bent, a crackling noise is heard (known as tin cry) due to the friction of the crystals as they ride over one another.

11.5 Applications

Tin coatings are most commonly produced by immersing the prepared work in a bath of molten tin, but they may also be producing by applying the tin in molten, stick or powdered form to the prepared and heated surface. The hot dipping process may be used to effect tinning and soldering simultaneously, as for example in the manufacture of motor car radiator blocks and refrigerator cooling units.

Tin coating may be applied to cast iron, steel, copper and its alloys such as brass and bronze. The production of a tin coating involves the formation of a thin layer of alloy or intermetallic compound between the outer layer of pure tin and the base metal. The alloy layer formed on iron or steel consists of the compound $FeSn_2$, while that on copper and its alloys consists of the compounds Cu_6Sn_5 and Cu_3Sn.

The most important outlet of tin is to provide hygienic protective coatings and give a pleasing appearance to articles made from stronger constructional basis metals. Tin-lead alloy coatings applied in a similar manner to that in use for pure tin find employment where non-toxicity is not essential and where the cheaper coating can safely be used.

A second important application of hot tinning is for soldering. Soldering has special interest to the electrical and electronic industries where reliable joints must be achieved by spot soldering at high speeds. A third purpose is to assist binding of another metal to the basis metal as in the tinning of bearing shells prior to lining with white metal.

Chapter 12
Magnesium

12.1 Introduction

Magnesium (Mg), the lightest metal, was first isolated by Sir Humphry Davy in 1808. In 1828 the French scientist Bussy reduced magnesium chloride with potassium, and Faraday first electrolyzed molten magnesium chloride to get magnesium metal in 1833. In 1852 Bunsen continued Faraday's work and produced some quantity of metallic magnesium. Modern electrolytic cells are designed on the principles of Faraday and Bunsen.

Magnesium is the only metal in group IIA of the periodic table which finds extensive application. The general uses of the metal fall under two main heads, namely nonstructural and structural. i) Nonstructural uses include application in alloying, deoxidation and desulphurization of molten metals, modifying the structure of graphite in cast irons. ii) Structural uses: alloys of Mg-Al have properties such as light weight, machinability, damping capacity, rigidity, and resistance to corrosion by chemicals.

12.2 Sources

Magnesium occurs in sizable amounts (2.35%) in the Earth's crust. An appreciable stock of magnesium is in sea water, which averages 0.3% $MgCl_2$, 0.04% $MgBr_2$, 0.18% $MgSO_4$. Although not as abundant as aluminium, magnesium ores are yet widespread. Magnesite is mined in Greece, Russia, Canada, USA, Austria and India. Dolomite, the double carbonate of magnesium and calcium is still more abundant, large quantities occurring in Europe and elsewhere. Brucite, containing a higher percentage of magnesium than any other mineral, is confined to somewhat

S.K. Dutta and D.R. Lodhari, *Extraction of Nuclear and Non-ferrous Metals*,
Topics in Mining, Metallurgy and Materials Engineering,
https://doi.org/10.1007/978-981-10-5172-2_12

Table 12.1 Important minerals of magnesium

Sr No	Mineral	Chemical Formula	Theoretical % of Mg
1.	Magnesite	$MgCO_3$	29
2.	Dolomite	$MgCO_3$, $CaCO_3$	13
3.	Sea Water	$MgCl_2$ and $MgSO_4$	0.13
4.	Carnallite	$MgCl_2.KCl.6H_2O$	9
5.	Brucite	$Mg(OH)_2$	42

localized areas. Salt beds, formed by the natural evaporation of water from inland seas, furnish magnesium salts which have been used as a source of magnesium. The most notable of these beds occurs at Stassfurt in Germany and contains many magnesium salts including carnallite. Table 12.1 shows the important minerals of magnesium.

Magnesium is successfully produced from sea water, which is the potential supply as raw material. Due to low concentration, magnesium recovery from sea water is costly and is employed in countries (USA, Norway, UK) where do not have other raw materials.

India has extensive deposits of magnesium in the form of magnesite ($MgCO_3$) at Chalk Hills of Salem district, in Tamil Nadu. The deposit is a high grade (above 46% magnesite). Apart from Salem, the other important magnesium deposits are located in southern Karnataka and Uttar Pradesh. Smaller deposits (some of these deposits are dolomite) are found in: Idar (Maharashtra), Kurnool (Andhra Pradesh), Tiruchirapalli (Tamil Nadu), Coorg (Karnataka), Dungarpur (Rajasthan), Singhbhum (Bihar). The Indian Ocean also contains on an average 0.6% $MgCl_2$.

12.3 Extraction of Magnesium

Ore treatment depends on the process of extraction of magnesium, to produce either MgO or $MgCl_2$ as follows:

(1) Pyrometallurgical process → to produce MgO.
(2) Electrometallurgical process → to produce $MgCl_2$.

Hydrometallurgical process is not used for extraction of magnesium because magnesium has good affinity for Cl^- and SO_4^{2+} ions, and hence it forms $MgCl_2$ and $MgSO_4$. Since $MgCl_2$ and $MgSO_4$ are soluble in water, they are already in solution form in sea water. Hence, hydrometallurgy is not a useful process for extraction of magnesium.

12.3.1 *Pyrometallurgical Process*

12.3.1.1 Method of Production of MgO

(a) Magnesite ore ($MgCO_3$) is first of all crushed and calcined at 600–700 °C:

$$MgCO_3 = MgO + CO_2 \quad (12.1)$$

(b) Dolomite ($MgCO_3.CaCO_3$) is crushed and calcined at 1350 °C, then calcine oxides are treated with water to get slurry of calcium and magnesium hydroxides, which are leached with sea water to precipitate [$Mg(OH)_2$]. The precipitate is filtered at 90 °C under vacuum and ignited to form MgO.

$$MgCO_3.CaCO_3 = MgO.CaO + 2CO_2 \quad (12.2)$$

$$MgO.CaO + 2H_2O = Mg(OH)_2 + Ca(OH)_2 \quad (12.3)$$

$$2Mg(OH)_2 + 2Ca(OH)_2 + [MgCl_2 + MgSO_4](\text{Sea Water}) = 4Mg(OH)_2 + CaCl_2 + CaSO_4 \quad (12.4)$$

$$Mg(OH)_2 = MgO + H_2O \quad (12.5)$$

This process is based on the principle that calcium has more tendencies to combine with Cl^- and SO_4^{2+} ions as compared to magnesium. Hence, leaching with sea water is done to precipitate $Mg(OH)_2$ out from the solution.

12.3.1.2 Thermal Process (Reduction of MgO)

It deals with reduction of MgO by a reducing agent. As per Ellingham diagram (Fig. 12.1), it is found that carbon, aluminium, or silicon can be selected as reducing agent, because at high temperatures they readily form very stable oxides due to their good affinities to oxygen than magnesium on high temperature. Main difficulty in extraction of magnesium is that magnesium is a very reactive metal at low temperatures; it readily forms very stable oxide (as shown in Fig. 12.1). In addition to high oxide stability necessitating very high temperatures for reduction, a further difficulty is encountered when an attempt is made to condense the resultant magnesium vapour; the reaction completely reverses (reaction 12.6), the overall effect is distillation of the oxide and carbon present in the original charge, carbon monoxide usually a reducing agent acting in this case as an oxidizing agent. It is the reversal of the reaction that had long been the main obstacle in the carbothermic reduction of magnesium.

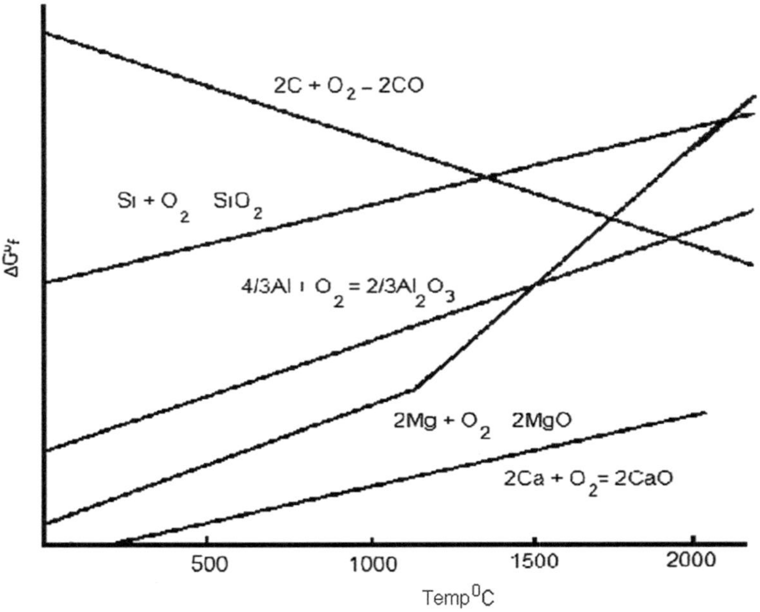

Fig. 12.1 Ellingham diagram for some oxides

From the Ellingham diagram (Fig. 12.1), it is found that: (a) at all temperatures, CaO is more stable than MgO; (b) at > 1550 °C, Al_2O_3 is more stable than MgO; (c) at > 1850 °C, CO is more stable than MgO; (d) at > 2350 °C, SiO_2 is more stable than MgO. Hence, Al, C, or Si can be reduced MgO above 1550 °C, above 1850 °C or above 2350 °C respectively.

(1) Reduced by Carbon: Carbon is the cheapest reducing material, reduction reaction take place at 1900 °C:

$$MgO\,(s) + C\,(s) \rightleftarrows Mg\,(g) + CO\,(g) \tag{12.6}$$

Here both the products are in a gaseous state (boiling point of magnesium is 1107 °C); hence, during cooling below 1400 °C, the backward reaction take place very fast. The reduced metal evaporates and has to be condensed from the gas phase at low temperature. This involves the risk of a retro-reaction or of the formation of carbide. To avoid this type of backward reaction (a) it should be cooled at 300 °C immediately or (b) by applying vacuum to reduce the partial pressure of magnesium vapour. The calculated lowest required reduction temperatures according to reaction (12.6) are shown in Table 12.2.

Reaction (12.6) takes place at 1900 °C, and according to Table 12.2 a vapour pressure of magnesium above 760 torr (1 atm) is therefore to be expected;

Table 12.2 Effect of temperature on the vapour pressure of magnesium during carbothmic reduction

Serial No.	Pressure (Torr)	Temperature (°C)
1.	760	1840
2.	100	1630
3.	10	1440
4.	1	1270

but difficulties may arise in the condensation, as the temperature of condensation is considerably below that for reduction. In this case intensive retro-reaction should be expected. By quenching with hydrogen or natural gas, metallic magnesium could anyway have been produced on that occasion, but this had to be subjected to a subsequent vacuum distillation.

More favourable conditions for reduction are obtained by reduction in a vacuum. From Table 12.2, it may be seen that reduction is possible even at 1400 °C. This considerably reduces the difficulties met with in the condensation. It appears that the reduction of MgO takes place and futher the product carbon dioxide gas reacts with excess carbon to form carbon monoxide.

$$MgO(s) + CO(g) = Mg(g) + CO_2(g) \tag{12.7}$$

$$CO_2(g) + C(s) = 2CO(g) \tag{12.8}$$

Reaction (12.8) seems to be the rate controlling step for the overall reaction (12.6). The optimum carbon monoxide gas pressure for 1400 °C is 0.1 torr. The reaction rate decreases with pressures both higher and lower than this. On the other hand, the rate of the pure solid-state reaction (12.6) rises with rising temperature.

The advantages of vacuum metallurgical reactions are as follows:

(a) By lowering the pressure of the gaseous reaction products, the concentration gradient is increased, so that the reaction rate may be increased.
(a) By pumping off gaseous reaction products, the equilibrium is disturbed in the direction of the desired reactions, so that the yield is increased.
(b) High vacuum reduces the extent of reactions between the metal vapours and the residual gases in the gas atmosphere.
(c) A reduction of the residual gas pressures to sufficiently low values results in maximum evaporation rates.
(d) The absence of an oxygen or nitrogen atmosphere increases the stability of a large number of metals i.e. for getting high purities of metal production.

(2) Reduced by Al: Since aluminium is comparatively costly, so this method is not economically feasible. Reduction takes place at 1650 °C in inert atmosphere.

$$3MgO\,(s) + 2Al\,(l) = 3Mg\,(g) + Al_2O_3(s) \tag{12.9}$$

(3) Reduced by Si: Since silicon can be easily available in the form of ferro-silicon (Fe–Si) which involves cheaper materials. Reduction takes place at 2400 °C:

$$2MgO\,(s) + Si\,(s) = 2Mg(g) + SiO_2(s) \tag{12.10}$$

Reduction temperature can be lowered by applying a vacuum to distill off the magnesium. If the process is carried out in a vacuum (i.e. at a pressure of 0.1 mm of Hg), then the temperature required for the reaction is considerably lower at 1100–1200 °C.

(4) Pidgeon process: Pidgeon and Alexander developed the Pidgeon process in 1944. It is based on the reduction of calcined dolomite (MgO.CaO) by Fe–Si in the presence of CaO and under vacuum.

$$2MgO.\,CaO\,(s) + Fe - Si\,(s) = 2Mg\,(g) + 2CaO.SiO_2(s) + Fe\,(s) \tag{12.11}$$

Ground fine size calcined dolomite (MgO.CaO) is briquetted with powdered Fe–Si (75% Si), 6:1 ratio; and reduction is carried out in retorts at 1200 °C and under vacuum of 0.1 mm of Hg. Small retorts (0.25 m diameter and 3 m length), made of Ni-Cr steel, are externally heated (by gas burners) and there are about 20 retorts in a furnace. Baffles are provided to prevent solid material going along with gaseous product and not for condensation of metal vapours.

12.3.1.3 Reaction Mechanism

It was originally believed that the Pidgeon process involved essentially solid–solid reaction. Therefore, finely ground materials were recommended to facilitate the reaction. However, it is now accepted that the reaction produces a liquid at some intermediate stage. The following distinct stages are observed:

(1) The initial reaction takes place between Fe–Si and CaO to produce a liquid Ca–Si–Fe alloy, which permeates the briquette and forms a metallic network. This reaction takes place rapidly at around 1000 °C and is mildly exothermic. The ternary alloy (Ca–Si–Fe) serves as the main reducing agent.
(2) Magnesium vapours are produced by the reduction of MgO by the Ca–Si–Fe alloy. At this stage, the pressure builds up rapidly, slowing down the rate is governed by the rate at which magnesium vapour can escape from the briquettes.

Another mechanism has been proposed by Toguri and Pidgeon (in 1962) that the presence of the SiO vapour in the Pidgeon process controls the kinetic of the reactions:

$$MgO\,(s) + Si\,(s) = Mg\,(g) + SiO\,(g) \tag{12.12}$$

$$2SiO\,(g) + 2CaO\,(s) = 2CaO.\,SiO_2(s) + Si\,(s) \tag{12.13}$$

12.3.2 *Electrometallurgical Process*

12.3.2.1 Method of Production of $MgCl_2$

There are several methods, but the following two are very popular:

(1) Dow process → from sea water.
(2) Reductive-Chlorinization of MgO → from MgO.

(1) Dow process: Sea water contains only 0.13% Mg in the form of $MgCl_2$ and $MgSO_4$. First of all evaporation of sea water is done, then a reaction with milk of lime to precipitate milk of magnesia [$Mg(OH)_2$]. The precipitate is filtered and treated with 10% HCl to form $MgCl_2$ solution which is evaporated to form pure crystal of $MgCl_2.6H_2O$.

$$[MgCl_2 + MgSO_4] + 2Ca(OH)_2 = 2Mg(OH)_2 + CaCl_2 + CaSO_4 \tag{12.14}$$

$$Mg(OH)_2 + 2HCl = MgCl_2 + 2H_2O \tag{12.15}$$

(2) Reductive-Chlorinization of MgO:

$$MgO + Cl_2 = MgCl_2 + 1/2O_2 \tag{12.16}$$

The above reaction is thermodynamically possible at temperature 250–300 °C and at 25–30% by volume concentration of chlorine gas. But the rate of reaction is very slow and so it cannot be accepted commercially. Hence, the above reaction must be carried out at temperature 500–700 °C and at high concentration of chlorine gas (80–90% by volume).

In electrolytic cell, $MgCl_2$ is used as electrolyte. The chlorine gas is evolved at the anode from $MgCl_2$ ($MgCl_2 \rightarrow Mg^{2+} + 2Cl^-$). That chlorine gas is utilized for chlorinization of MgO. But this chlorine gas is not pure, as it contains oxygen as impurity. Due to presence of oxygen in chlorine gas, this favours backward reaction, i.e. again formation of MgO. To prevent the backward reaction, temperature is increased to 800–1000 °C and some carbon (in form of coal or coke) is added to take care of oxygen. Now the actual reaction taking place is:

$$MgO(s) + C(s) + Cl_2(g) = MgCl_2(l) + CO(g) \tag{12.17}$$

$$MgO(s) + CO(g) + Cl_2(g) = MgCl_2(l) + CO_2(g) \tag{12.18}$$

Since chlorinization of MgO is carried out in presence of reducing agent (carbon), this process is known as the reductive chlorinization process. Melting point of $MgCl_2$ is 718 °C, so at the reaction temperature $MgCl_2$ is in molten condition.

12.3.2.2 Electrolytic Process

Extraction of magnesium metal was done by fused salt electrolysis of $MgCl_2$. It is not possible to electrolyze $MgCl_2$ in an aqueous solution, because: (a) magnesium is highly electro positive ($Mg = Mg^{2+} + 2e$, $E_0 = 2.37$ and $H_2 = 2H^+ + 2e$, $E_0 = 0.0$) and hence evolution of hydrogen gas starts at cathode (due to decomposition of water) before deposition of any magnesium metal; (b) magnesium is a very reactive metal and hence liberated magnesium vapour will react with oxygen (due to decomposition of water) to form MgO.

(1) Precautions should be taken during electrolysis of $MgCl_2$:

$$\text{During electrolysis of } MgCl_2\text{, the reaction is : } MgCl_2 \rightarrow Mg^{2+} + 2Cl^- \quad (12.19)$$

$$\text{At cathode : } Mg^{2+} + 2e \rightarrow Mg \quad (12.20)$$

$$\text{At anode : } 2Cl^- - 2e \rightarrow Cl_2 \quad (12.21)$$

The deposited magnesium metal and liberated chlorine gas should not come in contact with each other to form $MgCl_2$ once again. This is achieved by: (a) use of a hood; and (b) improving specific gravity and surface tension of the bath.

(2) Since magnesium is highly reactive, magnesium liberated at the cathode should be protected from outside atmosphere (i.e. oxygen); inert atmosphere is used in electrolysis.
(3) $MgCl_2$ should be anhydrous, otherwise it leads to difficulties due to: (a) steam evolution in the bath; (b) evolued chlorine gas contains HCl and water vapours; (c) increase in thermal voltage of the cell leading to difficulties in the cell operation.
(4) Theoretically, electrolysis of only $MgCl_2$ is possible, but not in practice. Because $MgCl_2$ has properties like: (a) poor electrical conductivity; (b) high melting point (718 °C); (c) low specific gravity; (d) very high viscosity and high surface tension.

Hence fluxes like KCl (15%), NaCl (30%), and $CaCl_2$ (45%) are added in $MgCl_2$ (10%) to improve all the above properties. All these fluxes are very stable and hence do not react with deposited magnesium. Sometimes, improving efficiency of the bath can be obtained by adding: molten carnallite (i.e. $MgCl_2$, KCl), and other additives like NaF, CaF_2 etc. Hood is used to: (a) prevent contact between magnesium and chlorine gas liberated at cathode and anode respectively; (b) protect cell from short circuiting by the floating molten magnesium metal.

The electrolyte bath shows the following desired properties:

(1) Good electrical conductivity.
(2) Low melting point of bath (e.g. melting points of $MgCl_2$, KCl, NaCl and $CaCl_2$ are 718 °C, 768 °C, 800° C and 774 °C respectively; combination of all, in above proportion, form low melting point of the mixture which melts at 550 ° C).
(3) High specific gravity of the bath (i.e. for easy removal of magnesium, it should float on the bath, hence specific gravity of the bath should be higher than magnesium metal).
(4) Low viscosity and surface tension of the bath; this also helps for easy floating of the magnesium metal on the bath.

Electrolytic cell for Mg production is shown in Fig. 12.2. Graphite rod and steel plates are used as anode and cathode (electrodes) in the electrolyte bath. The electrolyte bath temperature is maintained at about 670–750 °C. High current (30–50 kA) and low voltage (6.6 v) gives good heating effect. Molten magnesium is floating on the bath; from time to time, magnesium should be taken out from the bath to the ladles. Chlorine gas is liberated at anode, should be collected and send for the chlorinization of MgO to form $MgCl_2$. 99.5% purity of magnesium. Impurities amounting to 0.5% (metallic [those which get deposited along with magnesium are iron, silicon, aluminium, copper, nickel, sodium, patasium, calcium etc.] and non-metallic [$MgCl_2$, KCl, NaCl, $CaCl_2$ and MgO which comes along with metal during tapping to ladle]) are present in magnesium. This presence of impurities affects mechanical properties and corrosion resistance of magnesium metal.

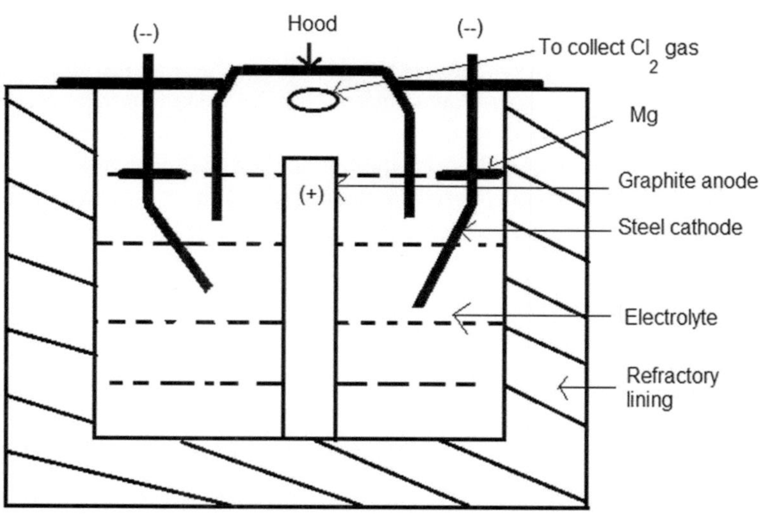

Fig. 12.2 Electrolytic cell for magnesium production

12.3.2.3 Refining of Electrolytic Magnesium

There are two methods for refining of electrolytic magnesium:

(1) Fusion with fluxes to remove non-metallic impurities.
(2) Retorting to remove metallic impurities.

(1) *Fusion with fluxes:* It removes only non-metallic impurities. The flux used is a mixture of $MgCl_2$, KCl, NaCl, $CaCl_2$ and MgO. The using flux serves two purposes: (a) it protects the metal from atmospheric oxidation during fusion; and (b) it removes impurities in the form of slag.

Firstly, the molten metal is thoroughly mixed with this flux by stirring and the whole charge is melted. Then molten bath is allowed to settle quickly at about 700 °C, until a solidified crust of slag forms on the surface. The slag crust is broken and refined magnesium is cast.

(2) *Retorting:* It removes metallic impurities by vaporization of magnesium in the retort kept under vacuum (0.1 mm of Hg pressure). Due to high vapour pressure of magnesium, it can be easily separated from impurities (like iron, silicon, aluminium, copper, nickel, calcium etc.) by vaporization. But some impurities like sodium, patasium etc. have vapour pressure very close to magnesium vapour. However, with close control of vacuum, magnesium can be separated even from these impurities (sodium, patasium etc.).

Apparatus consist of: (a) a sealed retort with its lower end in a heating furnace to serve as a vaporizer and the top end with water cooling to serve as condenser; (b) a vacuum is maintained in the retort. Also vacuum is maintained in the furnace to reduce oxidizing effect of air on the hot part of the retort and the coils; (c) shield is used to separate vaporizer and condenser part.

Electrolytic magnesium is heated in the retort in a vacuum (0.1 mm of Hg pressure) at 600 °C. At 450–500 °C magnesium vapour condenses in upper part of the retort in the form of bright crystals. Magnesium metal is purest in the middle portion; at the top part, it may be contaminated by more volatile impurities (sodium, patasium); at the bottom part, it may be contaminated by less volatile impurities (Al). Hence, magnesium crystals from the middle portion are separated from the rest, remelted (powder sulphur spread on the surface of melts to protect oxidation by SO_2 gas) and cast into desired form. Final purity obtained is about 99.99% Mg.

12.3.3 Other Processes for Extraction of Mg

These processes are used for small quantity production:

(1) Alumino-thermic reduction of MgO:

$$3MgO + 2Al = 3Mg + Al_2O_3 \tag{12.22}$$

(2) Reduction of $MgCl_2$ by Na:

$$MgCl_2 + 2Na = Mg + 2NaCl \tag{12.23}$$

(3) Fuse salt electrolysis of $MgSO_4$.
(4) Electrolysis of aqueous solution of $MgCl_2$ by using Hg cathode.

12.4 Properties

Magnesium is a silvery white metal and one of the lightest (specific gravity 1.74). The atomic weight of magnesium is 24.32, melting point and boiling point are 651 °C and 1107 °C respectively. It is divalent (Mg^{2+}) and has HCP structure. It is hard and less ductile in nature. In the form of thin ribbon or fine powder, magnesium burns in presence of air and evolving intense heat.

Being electro positive, magnesium reacts with acid evolving hydrogen gas. But it resists attack of HF, chromic acid and alkalis. When exposed to dry air, pure magnesium reacts and oxide film (MgO) forms on its surface and protects it from further oxidation. Hence, pure magnesium has good corrosion resistance property; but corrosion resistance decreases with increasing impurities content.

Like many other metals, pure magnesium has mechanical properties that are not very useful for engineering uses; and hence, for industrial applications, magnesium alloys are used. Magnesium is mainly alloyed with aluminium, zinc, manganese, zirconium etc. to improve its strength, ductility, corrosion resistance and smooth fabrication. Important alloys are: (a) Mg–Al (6–10% Al); (b) Mg–Zn–AL–Mn (3% Zn, 6% Al, 0.3% Mn); (c) Mg–Zn–Zr (4–5% Zn, 0.7%Zr).

12.5 Applications

Magnesium is widely used as an alloying element to improve the properties of alloys: (a) many Al base alloys containing less than 2% Mg, e.g. duralumin, piston alloys etc.; (b) in cast iron 0.1% Mg converts the brittle structure to a malleable one and thus improves its mechanical properties; (c) in some Cu base alloys, a small amount of magnesium improves strength and corrosion resistance; and (d) it is also used to improve the mechanical properties of lead base alloys containing calcium, tin and indium.

As a pure metal, magnesium is used in powder or ribbon forms for photographic purposes. It has also been in use as a deoxidizer and desulphurizer of the molten metals. Due to low specific gravity of Mg-alloys, they are used in aircraft industries.

Three promising fields have been opened up as follows:

(1) The excellent etching characteristics of magnesium have found use in the printing trade, several newspapers are printed directly from etched magnesium plates.
(2) A dry cell battery with a magnesium anode is said to have twice the capacity of an ordinary zinc battery.
(3) The high anodic character of magnesium makes it highly suitable for use as a sacrificial metal in the protection of iron and steel. Magnesium anodes are used for cathodic protection to pipelines, storage tanks and underground cables as sacrificial anode.

Chapter 13
Nickel

13.1 Introduction

Nickel (Ni) was obtained for the first time by the Swedish chemist A.F. Cronstedt, in 1751, who suggested the name for the element. Although nickel (Ni) is a common constituent of the Earth's crust being more widely distributed than copper, lead and zinc. Nickel has been found in economic quantities only in Canada, New Caledonia, Finland, Russia, Cuba, Griqualand, although other deposits are known. By far the most important of these areas is the Sudbury region of Ontario, Canada. The presence of nickel (and copper) in this region was known as early as 1856. In 1889, the Oxford Copper Co found a separation method for nickel and copper sulphides by taking advantage of their differing solubilities in sodium sulphide. The copper sulphide is readily soluble and nickel sulphide is practically insoluble.

13.2 Sources

Nickel (Ni) is mostly associated with copper and iron compounds in its ore. In nature nickel occurs as oxides, sulphides and silicates forms. The principal ore minerals of nickel are the (I) sulphides: (a) millerite (NiS) and (b) pentlandite [$(NiFe)_{11}S_{10}$], it is a nickeliferrous pyrrhotite mixed with chalcopyrite (Cu_2S, Fe_2S_3) and rocky gangue; (II) hydrous silicate: garnierite [$H_2(Mg,Ni)SiO_4$]; (III) mixed copper-nickel ore (with nearly equal amount of Cu); (IV) oxide: nickeliferous latenites or Ni-oxide ore; and (V) sea nodules (oxide).

Large deposits are available in Canada and biggest producer of nickel in the world is International Nickel Co (INCO), Canada. INCO contributes for nearly 80% of world nickel production. Total requirement of nickel in India is fully imported.

Apart from above, several tonnes of nodules are scattered across the ocean floor. These nodules, whose principal constituents are Mn (23.86%), Fe (13.8%),

S.K. Dutta and D.R. Lodhari, *Extraction of Nuclear and Non-ferrous Metals*,
Topics in Mining, Metallurgy and Materials Engineering,
https://doi.org/10.1007/978-981-10-5172-2_13

Al (2.86%), Mg (1.66%), Ni (0.98%), Cu (0.52%), Co (0.35%) and siliceous; are collectively termed as *manganese nodules.* In terms of their nickel content, they can also be called *nickel nodules*, since they are a potential source of nickel.

13.3 Extraction of Nickel

Extraction of nickel is more or less similar to that of copper extraction. Basically extraction of nickel from Cu–Ni ore which involves basically three stepes:

1. Concentration,
2. Treatment of Ni–Cu sulphide concentrate, and
3. Refining.

13.3.1 Concentration

Flow diagram of nickel extraction is shown in Fig. 13.1. Cu–Ni ore (1.3% Cu, 1.2% Ni) is concentrated by flotation to get bulk Ni–Cu concentrate (7% Cu, 6% Ni), and nickel-ferrous sulphide concentrate (0.9% Ni) as tailing. Again by flotation separating of Ni–Cu sulphide concentrate (10% Ni, 2% Cu, 40% Fe, 30% S) and Cu concentrate (30% Cu, 1% Ni) are done; Cu is recovered from Cu concentrate.

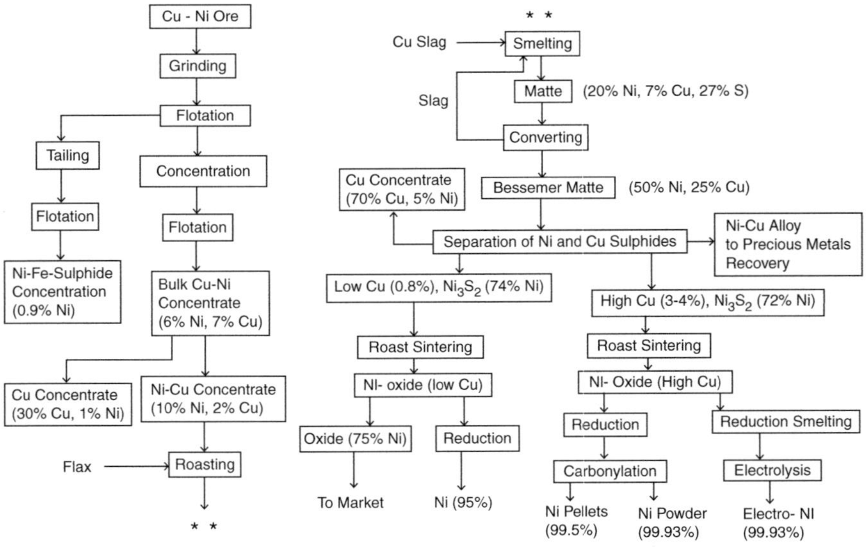

Fig. 13.1 Flow diagram of extraction of nickel

13.3.2 *Treatment of Ni–Cu Sulphide Concentrate*

Ni–Cu concentrate consists of the sulphides of these metals together with iron pyrite (FeS_2), silica (SiO_2) and small amounts of cobalt (Co), selenium (Se), tellurium (Te), gold (Au), silver (Ag) and metals of platinum (Pt) group. Hence, the treatment is designed to remove impurity and to concentrate nickel, copper and precious metals into matte for further treatment. The treatment consists of the following steps:

(i) Controlled roasting,
(ii) Smelting,
(iii) Converting,
(iv) Separation of nickel and copper sulphides,
(v) Roast sintering, and
(vi) Reduction.

13.3.2.1 Roasting

Nickel has to be separated from the iron present in the sulphide concentrate; this is similar to copper metallurgy. The concentrate is first partially roasted to selective oxidized of FeS_2. This is done either in multiple hearth roasters or in a fluidized bed roaster. The later is preferred because of its high output and the rich sulphur dioxide (SO_2) gas stream it generates. During controlled roasting, in the temperature range of 550–600 °C, about 40% of the sulphur is oxidized and enough heat is generated to make the roasting in the fluid bed autogenous process.

13.3.2.2 Smelting

The roasted calcine, which also contains the desired amount of siliceous flux, is smelted in a reverberatory furnace to produce a matte containing nickel and copper sulphides (20% Ni, 7% Cu, 40% Fe, 27% S) and slag containing the gangue and iron oxide (0.3–0.4% Ni, 0.2% Cu, 39% Fe, 35% SiO_2). The converter slags from both nickel and copper converters are charged to the reverberatory furnace. The molten matte is tapped into the ladles and taken to converters.

13.3.2.3 Converting

The matte and flux (SiO_2) is charged in a converter and blowing is done to produce Ni-enriched matte. The blowing is not carried to the metal stage but serves only to oxidize FeS_2, the formation of a small amount only of metal is allowed to enable this to act as a collector for the precious metals.

The converting is completed at about 1150 °C to produce Bessemer matte (50% Ni, 25% Cu, 0.7% Fe, 21.5% S). The slag (2% Ni, 1.5% Cu, 40% Fe and 25% SiO_2) is returned to the smelting furnace for recovery of nickel and copper.

13.3.2.4 Separation of Ni and Cu Sulphides

Bessemer matte is subjected to a slow cooling process wherein the matte is cooled from its melting point to about 400 °C in a period of 3 days. During slow cooling, first Cu_2S precipitates and grows. At 700 °C, a second phase, namely, a metallic Ni–Cu alloy phase (64% Ni, 16% Cu, 10% S), starts to precipitate. At 575 °C, the third solid phase, Ni_3S_2, starts to precipitate. The slow cooling rate provides the necessary grain growth, so that in subsequent processing the sulphides can be recovered by flotation.

The slow cooled matte is grinded. Before flotation, the Ni–Cu alloy (which contains >95% precious metals) is separated by magnetic separation and sent for the recovery of precious metals. The copper sulphide concentrate produced by flotation, which contains about 70% Cu, 5% Ni and 20% S, is sent to the copper circuit. Nickel sulphite is recovered as a low copper, nickel sulphide (74% Ni, 0.8% Cu, 0.8% Fe and 22% S) and as high copper, nickel sulphide (72% Ni, 3–4% Cu, 0.8% Fe and 21% S).

13.3.2.5 Roast Sintering

Both the nickel sulphide fractions are roasted in fluidized bed roasters at the temperature range 1100–1250 °C to produce granular nickel oxide. The low copper, nickel oxide is marketed directly as nickel oxide or reduced to metal. The high copper, nickel oxide is sent to reduction and refining.

13.3.2.6 Reduction

For low copper, sintered nickel oxide is reduced by H_2 at 350–400 °C:

$$NiO + H_2 = Ni + H_2O \tag{13.1}$$

13.3.3 *Refining*

The sinter nickel oxide, containing 75% Ni, can be used as such in the production of steel; but for the production of metallic nickel the oxide has to be reduced and then refined. There are two methods: (1) The Mond process (or Carbonyl process) and (2) Electrolytic process.

13.3.3.1 The Mond Process

(a) Carbonyl Process

This process is based on the property that nickel possesses of combining with carbon monoxide to form a volatile compound, nickel carbonyl [$Ni(CO)_4$] which is stable below 60 °C but on heating to 180 °C breaks into the original forms.

As the nickel oxide sinter contains small amounts of sulphur and metallic elements which reduce the efficiency of the subsequent processing. So it is first given a preliminary oxidized roasting. Temperature of 750 °C is maintained and a flow of air admitted to oxidize the sulphur and metals.

$$Cu_2S,\ Ni_3S_2 + 5.5O_2 = 2CuO + 3NiO + 3SO_2 \tag{13.2}$$

Then calcine, high copper, nickel oxide, is leached with dilute H_2SO_4, the leaching removes large part of the copper as copper sulphate and left a residue of NiO.

$$CuO + H_2SO_4 = CuSO_4 + H_2O \tag{13.3}$$

This copper sulphate is filtered to remove the undissolved nickel oxide, the copper sulphate recovered by crystallization.

NiO is reduced to finely divide metallic nickel by an atmosphere of water gas (36% CO and 52% H_2) at 350–400 °C. To ensure efficient reduction two reducers are operated in series, the partially reduced material is conveying from the bottom of the first to the top of the second. The passage through each reducer takes about 4 h.

$$NiO + H_2 = Ni + H_2O \tag{13.4}$$

At the comparatively low temperature carbon monoxide plays very little part in the reduction, this is mainly carried out by the hydrogen. The result of the greater activity of hydrogen [reaction (13.4)] is to yield an exit gas which after removal of the water is very rich in carbon monoxide and thus suitable for the formation of nickel carbonyl in the next operation.

It is then cooled, passed on to volatilizer, where at a temperature of about 40–90 °C, the active nickel is exposed to a flow of carbon monoxide gas to form gaseous nickel carbonyl [$Ni(CO)_4$], which is carried out in the gas stream. The material passes through eight volatilizers in series; the process takes place in several days. This is a slow process.

$$Ni + 4CO = Ni(CO)_4 \tag{13.5}$$

This gas [$Ni(CO)_4$] is then passed through a tower containing nickel shots (4 – 6 mm diameter) at 180 °C and the compound nickel carbonyl is decomposed into metallic nickel and carbon monoxide (because nickel carbonyl is stable below 60 °C).

$$Ni(CO)_4 = Ni + 4CO \tag{13.6}$$

Close control of temperature is necessary because at 200 °C some carbon and carbon dioxide are formed by decomposing carbon monoxide in presence of nickel, which act as a catalyst. This carbon is absorbed by metallic nickel.

$$2CO = C + CO_2 \tag{13.7}$$

Decomposed metallic nickels are deposited on the nickel shot and increase its size to form nickel pellet. Time taken for pure nickel pellet to build-up, 10 mm diameter, is approximately 3 months. This nickel, containing <0.06% Fe and <0.09% Cu as impurities, is considered as pure metal.

(b) INCO Pressure Carbonylation Process

An elevated pressure of about 70 atmospheres is used. Higher pressure aids the reaction and permits higher operating temperature (180 °C). Under this conditions, carbonyls of iron, nickel and cobalt [$Fe(CO)_5$, $Ni(CO)_4$ and $Co_4(CO)_{12}$] are formed. $Ni(CO)_4$ is recovered by fractional distillation and later converted to metallic nickel in a pellet decomposer or a powder decomposer.

13.3.3.2 Electrolytic Process

Sintered Ni oxide is first reduced to metal by reduction smelting in presence of carbon and Ni–Si flux (assisting deoxidation):

$$NiO + C = Ni + CO \tag{13.8}$$

Crude nickel metal is then cast into anodes, which contain 95% Ni, 2% Cu, 0.75% Fe, and 0.7% S. Cathodes are produced in a somewhat similar to that are used in the preparation of electrolytic copper starting sheets. Thin nickel sheets are produced in starter cells.

Electrolysis is carried out in concrete tanks, solution of nickel sulphate and chloride containing about 60 gm of Ni^{2+} per litre, 95 gm of SO_4^{2-} per litre, 35 gm of Na^+ per litre, 55 gm of Cl^- per litre and 16 gm of H_3BO_3 per litre are used as electrolyte. Electrolysis is carried out at 60 °C at a current density of 170 A/m^2. On passage of current, nickel is dissolved from the anodes and is deposited on the cathode starting sheet. To prevent the deposition of copper and iron, the cathodes are enclosed in a canvas cover, which is tightly woven in order that a hydrostatic head of 25–50 mm is maintained over the level in the tank; thus ensuring a constant outward flow of electrolyte through the canvas (as shown in Fig. 13.2).

Actually nickel ions, set free at the anode, are not carried directly to the cathode but along with the copper and iron ions in solution flow continuously out of the tank

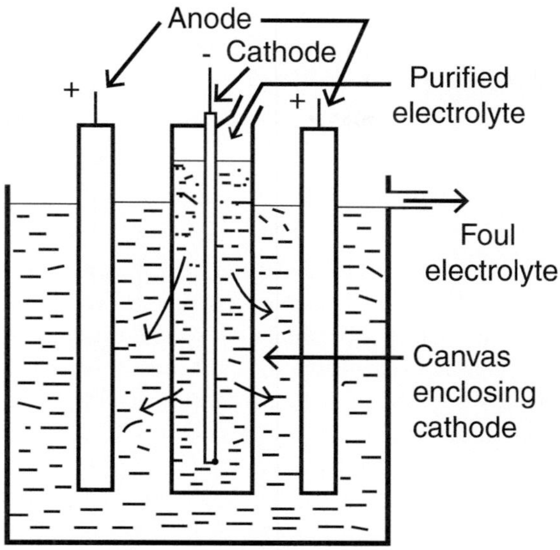

Fig. 13.2 Electro-refining of nickel

as foul electrolyte. The foul electrolyte is free from copper by agitation with active nickel powder, the copper is cementing out. Vigorous agitation is done by passing air, oxidizes the iron which is precipitates as $Fe(OH)_3$. Small amount of chlorine gas is passed to ensure oxidation to form precipitates of $Fe(OH)_3$. Cobalt is removed as cobalt hydroxide by further chlorine gas oxidation. At this stage cobalt is removed; further removal of impurities, such as arsenic and lead, are taking placed. The purified solution (contains 0.01 gm of Co per litre, 0.001 gm of Cu per litre, 0.004 gm of Fe per litre, 0.0002 gm of As and Pb each per litre) is then passed to the cathode compartment. When deposition is complete the cathodes are withdrawn, washed and either marketed in sheet form or melted and cast into ingots and shots (99.93% Ni, 0.05% Co, 0.005% Cu, 0.002% Fe, 0.001% S, 0.0002% As and 0.001% Pb).

13.4 Properties

The atomic weight of nickel is 58.69 and specific gravity of 8.88, melting point and boiling point are 1445 °C and 3075 °C respectively. Nickel is a divalent, non-toxic metal, has FCC structure; Nickel is strong and ductile metal and has high corrosion and heat resistance metal (important property). It is ferro-magnetic in nature. Nickel can be readily cold or hot worked, welded and machined. It retains its strength to an excellent degree at high temperatures and its ductility and toughness as well as strength at sub-zero temperatures. Another feature is the retention of its mechanical properties at high temperature. Nickel is dissolved in acids but it does not react with alkali (important property) i.e. high corrosion resistance to alkalies. It is one of the important alloying elements. It forms binary and ternary alloys with many metals which are used for resistance to corrosion and heat.

13.5 Applications

Nickel in its commercial pure (99.5), wrought and cast forms has wide and important industrial application due to good mechanical and special physical properties which it combines with good resistance to corrosion. Nickel is non-toxic, so it is used extensively in manufacture of food handling equipments. The ductility of nickel is also of a high order especially in the hot rolled condition, and the combination of strength and ductility, it can used in many constructional and mechanical applications.

Nickel and chromium are two strategically important alloying elements that are being increasingly consumed in the fields of chemical processing, space research, and nuclear reactor etc. Many uses of nickel are dependent on the combination of the properties of strength and ductility, which are the metal possesses. Nickel is also a valuable alloying constituent in more than 3000 commercial alloys which find wide applications, principally for their mechanical properties and resistance to corrosion and heat. Pure nickel acts as catalyst in many reactions: (a) cracking of natural gas ($CH_4 \rightarrow CO + H_2$); and (b) hydrogenation of vegetable oil.

Stainless steels, which contain a high percent of Ni (8%) and Cr, are known for outstanding resistance to high temperature oxidation and strong chemicals such as alkalis. Ductile nickel alloyed steels (up to 5.25% Ni) are used in various structural applications. Electric heaters and thermocouples use heat resistance alloys made of Ni, Fe and Cr. Due to non-toxic nature, nickel is used extensively in the manufacture of food handling equipment and in plant for the processing of pharmaceutical products.

Nickel forms so many non-ferrous alloys:

1. Inconel (80% Ni, 13% Cr and 7% Fe), resistance to corrosion at high temperature; it is used in petroleum refining, dairy, fruit juice, alkali, edible oil industries.
2. Monel (67% Ni, 30% Cu with Mn, Fe) for its good corrosion resistance and strength; it can be used in food, chemical, petroleum industries, hospital and kitchen equipment, marine service etc.
3. Cupronickel (30% Ni, 70% Cu), corrosion resistance, it can be used as heat exchanger tubes, steam turbine blades, coinage.
4. German silver (10–30% Ni, 5–30% Zn, 1–10% Pb and rest Cu), workability, corrosion resistance; ornamentation, jewellery, food equipment, musical and dental instruments.

Part III
Extractive Metallurgy of Less Common Metals and Ferro-Alloying Metals

The ferro-alloys may be defined as "alloys of iron with high percentage (>50%) of some particular elements other than carbon; they are used as alloying elements in iron and steelmaking". Their major field of application is as addition agents to molten steel and cast iron, and they function as: (a) deoxidizers; (b) alloying agents; or (c) graphitizers. A number of non-ferrous metals derived from oxide ores are basically used as alloying elements in steel making. This group includes chromium, manganese, silicon, molybdenum, tungsten, vanadium, and niobium.

Ferro-alloys are produced by reducing metals from their oxides. The starting materials are ores or ore concentrates. Fe-Si, Fe-Mn and Fe-Cr are produced from ores; Fe-W, Fe-Mo, Fe-V, Fe-Ti, and Fe-Nb are produced from concentrates. Reduction reactions may be characterized by stability of oxides formed at high temperatures. Figure III.1 shows standard values of free energy of oxide formation as a function of temperature, related to one mol of oxygen; this Figure III.1 is known as the Ellingham diagram. The stability of all oxides decreases with increasing temperatures, an exception being only carbon monoxide which becomes more stable with increasing temperature. The Ellingham diagram can be used to estimate the possibility of using an element as a reductant. An element, which forms a stronger oxide, can be used as reductant of a less strong oxide. A suitable reductant should be selected and conditions should be formed so as to ensure a high recovery of the valuable elements from the starting material. The most common reductants in the production of ferro-alloys are carbon, silicon and aluminium. As per the Ellingham diagram, it is found that carbon, aluminium or silicon can be selected as reducing agent, because at high temperatures they readily form very stable oxides. Carbon is the cheapest among them and because of that it is used in the production of Fe-Si, high C-Fe-Mn and high C-Fe-Cr. Iron, manganese, chromium, molybdenum, vanadium, titanium and other elements are formed carbides; hence for producing Fe-Cr or Fe-Mn with carbon as reductant, the ferro-alloys produced have a high content of carbon. For making low carbon ferro-alloys, other than carbon is used as reductant.

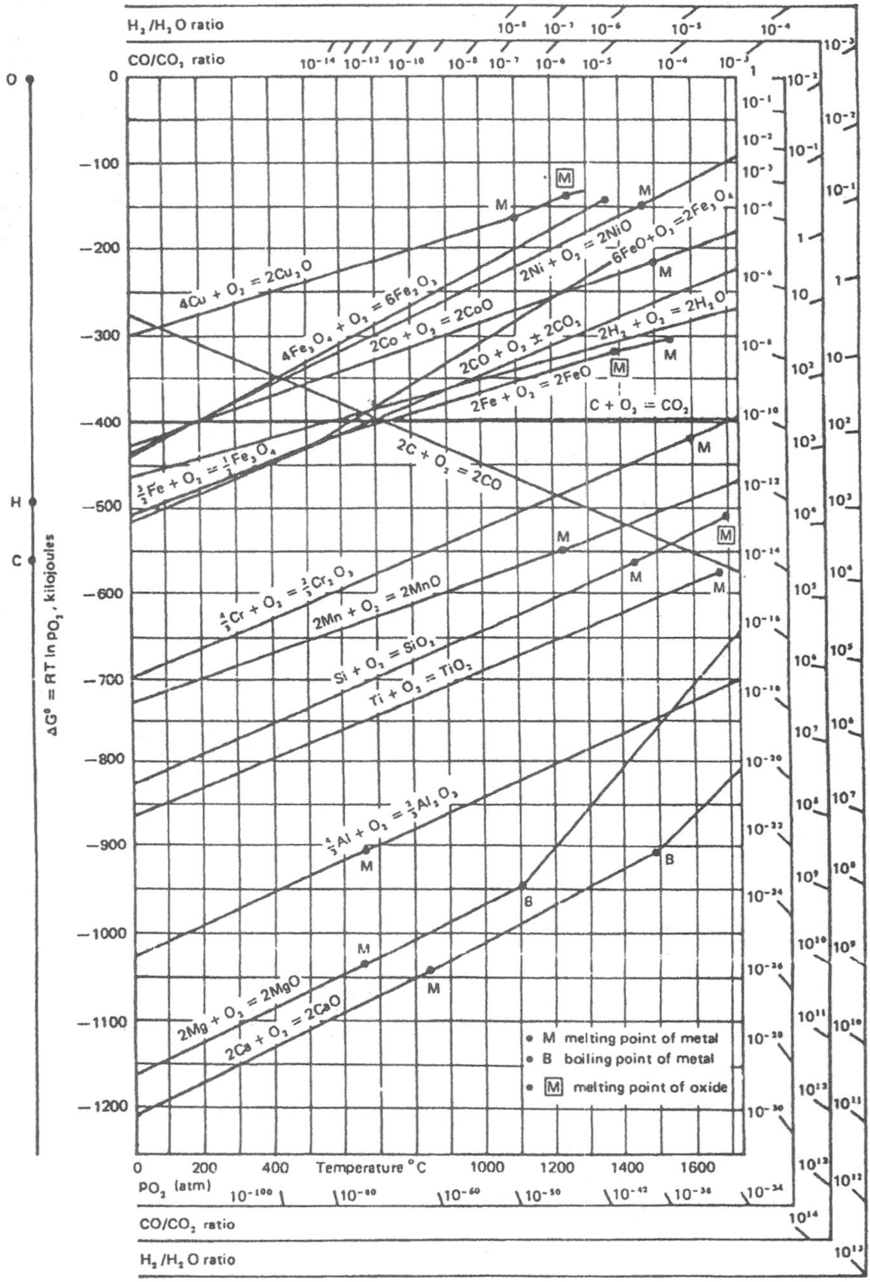

Fig. III.1 Ellingham diagram for metal oxides

In the production of ferro-alloys in submerged arc furnaces, the primary function of the carbonaceous material (mainly coke and coal) is to act as reductant: to react with the metal oxide to form metal and carbon monoxide:

$$MO + C = M + CO$$

It appears that the solid reductant undergoes several quite different reactions during ferro-alloy production:

(a) Boudouard reaction ($CO_2 + C = 2CO$) (e.g. ferro-manganese and ferro-chromium).
(b) Dissolution in the metal (e.g. ferro-manganese and ferro-chromium).
(c) React with SiO ($SiO + C = Si + CO$) (e.g. ferro-silicon and likely ferro-chromium).

Chapter 14
Silicon

14.1 Introduction

The history of silicon (Si) began shortly after 1800 when Davy concluded that silica was a compound, not an element. Silicon is second only to oxygen in occurrence in nature. It constitutes 27.6% of the Earth's crust. Man knows more than 200 different varieties of silicon: quartz, quartzite, chalcedony, rock crystal, opal, sand and many others. Silicon is a strong deoxidizer; this property determines basically its use in metallurgy. Commercially, pure silicon is used for the production of semiconductor, silicon bronzes, etc. It is used as a deoxidizer, in the forms of ferroalloy, in the production of all kind of steels. High amounts of ferrosilicon in the form of powder are used as a slag deoxidizer; it is also used for the reduction of various oxides in ferroalloy production.

14.2 Sources

Silicon occurs widely over the Earth's surface; minerals having a high content in silica, such as quartz, quartzite and chalcedony. Quartz is a compact mineral with crystal structure, specific gravity of 2.59–2.65, hardness 7, mostly colourless, white, grey or reddish, depending on impurities. Quartz is a relatively costly mineral and is used for the manufacture of crystalline silicon. SiO_2 content in quartz is 98% and over. Iron oxide in quartz should not exceed 0.3%.

Quartzite is a rock composed of quartz grains, cemented by a substance mainly containing silicon. Quartzites are widely distributed and their reserves are inexhaustible. Good grades of quartzite contain 96–97% SiO_2, 1% Al_2O_3, roughly 1% CaO and MgO, and not more than 0.02% P_2O_5. They do not require upgrading, and are readily available.

S.K. Dutta and D.R. Lodhari, *Extraction of Nuclear and Non-ferrous Metals*,
Topics in Mining, Metallurgy and Materials Engineering,
https://doi.org/10.1007/978-981-10-5172-2_14

Chalcedony is a thin fibrous, sometimes porous, mineral, of various colours. Its behaviour in metallurgical processes is somewhat worse than that of quartz and quartzite as it contains a lesser amount of SiO_2 (about 95%); its use is limited.

14.3 Extraction

14.3.1 Metallic Silicon

The quartzite is reduced by charcoal in an electric furnace to produce silicon.

$$SiO_2 + 2C = 2Si + 2CO \tag{14.1}$$

$$SiO_2 + 3C = SiC + 2CO \tag{14.2}$$

Any SiC formed is removed by secondary smelting with quartzite:

$$2SiC + SiO_2 = 3Si + 2CO \tag{14.3}$$

In the production of silicon or ferrosilicon, it appears that the major gaseous reagent is silicon monoxide (SiO), which can form through the reaction of silicon or silicon carbide with SiO_2 in the high-temperature region of the furnace. The silicon monoxide (SiO) product rises through the furnace and reacts with the reductant in the upper part of the furnace, to yield silicon carbide and carbon monoxide as products:

$$Si + SiO_2 = 2SiO \tag{14.4}$$

or

$$SiC + 2SiO_2 = 3SiO + CO \tag{14.5}$$

By production of SiO gas in the higher temperature (1650 °C) region, the reduction of SiO to silicon or silicon carbide is higher in the furnace:

$$SiO + C = Si + CO \tag{14.6}$$

or

$$SiO + 2C = SiC + CO \tag{14.7}$$

This silicon is suitable for metallurgical applications. However, silicon of a very high purity is required for semiconductor applications. Pure silicon is produced by first obtaining $SiCl_4$ gas through the chlorination of silicon, then purifying the $SiCl_4$

gas by distillation, which is followed by reduction with either zinc vapour or hydrogen gas at 1000 °C.

$$Si + 2Cl_2 = SiCl_4 \quad (14.8)$$

$$SiCl_4 + 2Zn = Si + 2ZnCl_2 \quad (14.9)$$

$$SiCl_4 + 2H_2 = Si + 4HCl \quad (14.10)$$

If the reduction is carried out in the presence of a heated silicon rod, the diameter of the rod gradually increases. This silicon rod is zone refined and used to grow single crystals, which are widely used in semiconductor applications.

14.3.2 Ferro-Silicon

The ore components of the charge for making ferro-silicon are minerals having a high content in silica, such as quartz and quartzite. Good grade of quartzite should have the lower content of alumina; the smaller would be the volume of slag, and therefore the lower the loss of electrical energy. Quartzite with a high content in alumina should be washed after grinding. The optimum size of quartzite lumps is 20–80 mm. Quartzite used in the production of ferro-silicon should meet the following requirements:

(1) Silicon content should be minimum 96%, desirably 97–99%.
(2) Amount of slag forming gangue materials (i.e. alumina [1%], magnesium oxide and calcium oxide [1%]) should be minimum.
(3) P_2O_5 content should not exceed 0.02%.
(4) Its moisture absorption should not exceed 5%.
(5) It should have no clay content.

Various carbon-bearing materials may be used as reducing agents in ferro-silicon production: carcoal, petroleum and pitch cokes, metallurgical coke, and coal. The main requirements to be met by a reducer are a low ash, low volatiles content, high strength of lumps, high electric resistance and cheapness. The best reducers are charcoal and petroleum coke; they contain little ash (1–2%), but are rather expensive. The ash of coke is the main source of aluminium in alloys. Therefore, low ash coke is needed as reducer in order to make 75% ferro-silicon with a low content in aluminium, as required. A large content of fines in coke breeze can hamper the evolution of gases and disturb the normal operation of the furnace. The presence of large lumps is likewise undesirable, since this can increase the electrical conductivity of the charge, which will necessitate rising of the electrodes and thus reduce the furnace productivity.

Ferro-silicon is commonly produced in the electric arc furnace from quartzite, coke and steel scrap (turnings of carbon steel having a low content of phosphorus).

The physical state of the charge is of prime importance for successful operation of a furnace. The charge materials must have constant moisture and lumps of coke breeze and quartzite should have only slightly varying size. A typical charge for the production of 75% grade of ferro-silicon is 2000 kg of quartz, 1000 kg of coke and 225 kg of steel turnings. Power consumption is about 10,000 kWh per tonne of ferro-silicon.

The reduction reaction of silicon from quartzite occurs with solid carbon:

$$SiO_2(1) + 2C(s) = Si(1) + 2CO(g) \tag{14.11}$$

With high temperature reduction which is typical of the process of smelting ferro-silicon, the process of reduction of silica into silicon at atmospheric pressure is most likely to occur in two stages:

$$SiO_2(1) + Si(1) = 2SiO(g) \tag{14.12}$$

$$SiO(g) + C(s) = Si(1) + CO(g) \tag{14.13}$$

Samples taken from lower levels in the furnace usually contain much silicon carbide (SiC). The formation of silicon carbide from the elements can only occur with large kinetic difficulties and requires a high mobility of atoms, which can only be achieved at a temperature above 1700 °C. On the other hand, the reaction is probable occur as:

$$SiO(g) + 2C(s) = SiC(g) + CO(g) \tag{14.14}$$

Solid inclusions of silicon carbide, if present, in the slag, can impair the fluidity of already tough siliceous slag. At corresponding temperatures, silicon carbide can be destroyed by metals and oxides, its destruction by iron following the reaction:

$$SiC(s) + Fe(1) = FeSi(1) + C(s) \tag{14.15}$$

At high temperature and in the presence of a solvent (iron with silicon) the aluminium oxide and calcium oxide, if present in the charge, are reduced by carbon and silicon:

$$Al_2O_3(s) + C/Si \rightarrow Al(1) + CO(g)/SiO_2(s) \tag{14.16}$$

$$CaO(s) + C/Si \rightarrow Ca(1) + CO(g)/SiO_2(s) \tag{14.17}$$

Industrial grade of ferro-silicon can thus contain up to 2% Al and up to 1.5% Ca.

14.4 Properties

Pure silicon obtained is a brown highly hygroscopic powder. Macro-crystalline silicon is obtained when silicon is melted and cooled; its colour is grey steel with metallic lustre. It is hard and brittle. Silicon is a metalloid with an atomic mass of 28.09, and specific gravity of 2.37. Melting and boiling points of silicon are 1414 °C and 2287 °C respectively. In its electric properties, silicon is a semiconductor material.

An addition of 1.3–2.0% Si to steel increases its hardness, ultimate strength, elastic and yield point and oxidation resistance of steel, but lowers the ductility.

Silicon can fuse together with iron in any proportion and forms a series of iron silicides. A number of chemical compounds are formed: ε-FeSi (33.3% Si), η-Fe_3Si_2 (25.1% Si), and ζ-Fe_2Si_5 (55.68% Si). The specific gravity of iron-silicon alloys varies with their silicon content. The melting point of standard grades of ferro-silicon does not exceed 1370 °C, 40–47% silicon alloy melts at 1260–1370 °C, while a 74–80% silicon alloy melts at 1320–1340 °C.

Silicon is a strong deoxidizer. Silicon reacts with oxygen to form silica (SiO_2), whose melting point is 1710 °C. Silicon and carbon form silicon carbide (carborundum, SiC), having a melting point above 2700 °C.

14.5 Applications

Metallic silicon is used mainly for alloying with copper, aluminium, magnesium and iron. Silicon is also used as a reductant in metallo-thermic reduction. Pure silicon (98–99%) is used for semiconductor applications and as silicon chips in the electronic industry.

Ferro-silicon is used for deoxidation and alloying of steel. Siliceous steels are widely used for making springs and like parts. Siliceous transformer iron (with 4% Si) is known to have relatively low power losses. The content of silicon may be up to 0.35% in tool steels, up to 0.37% in structural steels, and up to 5% in alloys grades. Ferro-silicon is used as a slag deoxidizer and for the reduction of various oxides in ferro-alloy production; e.g. in the production of Fe–Mo, Fe–V etc.

Chapter 15
Manganese

15.1 Introduction

Manganese (Mn) is an element of the 7th Group of the Periodic Table (Fig. 1.1). The earth's crust contains 0.09% Mn which is 12th in occurrence among the elements. Manganese is not found in elementary state in nature and occurs only in the form of compounds with other elements; these are mainly oxides, carbonates and silicates.

Manganese was first recognized as an element by the Swedish chemist Scheele, while working with the manganese dioxide ore pyrolusite, and was isolated by Gahn in 1774. Frequently found in conjunction with iron ores, the metal was named for the magnetic properties exhibited by pyrolusite from the Latin, *magnes*, or magnet; the German equivalent is *mangan* and the French, *manganese*.

The addition of manganese, as spiegeleisen, was initiated in 1856 by Robert Mushet in the Bessemer steelmaking process. In 1888 Robert Hadfield discovered the 14% manganese steels. Manganese alloys are the predominant constituent is manganese. The most important is ferro-manganese, an alloy of iron and manganese, containing about 80% Mn.

15.2 Sources

The ores of manganese are the oxides such as pyrolusite (MnO_2), which is the most important. Manganite (Mn_2O_3), braunite ($3Mn_2O_3.MnSiO_3$) and hydrated oxide are psilomelane. The pyrolusite (MnO_2) is the most important ore. The main impurities are alumina, silica and iron oxide. The ores of manganese can be divided into four categories determined by grade and quality, as follows:

(1) Exceptionally high grade ore characterized by a high content of manganese dioxide. This is known as battery ore, because of its use in the battery industry.

S.K. Dutta and D.R. Lodhari, *Extraction of Nuclear and Non-ferrous Metals*,
Topics in Mining, Metallurgy and Materials Engineering,
https://doi.org/10.1007/978-981-10-5172-2_15

(2) At the other extreme is manganiferous ore, usually containing 12% or less Mn and being used directly for making manganese pig iron.
(3) Ores containing not less than 45% Mn with a Mn:Fe ratio of 7:1. These are utilized for production of the ferro-alloy.
(4) Ore in which Mn:Fe ratio is too low, for production of ferro-alloy, that can be used for spiegeleisen.

Apart from the above, several tonnes of nodules are scattered across the ocean floor. These nodules, whose principal constituents are Mn (23.86%), Fe (13.8%), Al (2.86%), Mg (1.66%), Ni (0.98%), Cu (0.52%), Co (0.35%) and siliceous; are collectively termed as *manganese nodules*. In terms of their nickel content, they can also be called *nickel nodules*, since they are a potential source of nickel.

15.3 Extraction

15.3.1 Beneficiation

The manganese occurs as silicate, carbonate and oxide, and is so very finely disseminated that ultra-fine grinding is necessary to liberate the mineral values. Gravity concentration with heavy liquids or tables increases products' content only 1% or 2% higher in manganese than the charged material, there being no appreciable elimination of the major impurities, silica and iron oxide. Roasting and magnetic separation are slightly better than gravity separation. Non-magnetic concentrate contains 16–23% Mn, but recovery is only 40–50% of the total Mn in the ore. The magnetic fraction contains 5–14% Mn and 40–50% Fe with 0.6–1.0% P.

The low-grade manganese ores are given a reducing roast and leached with nitric acid, iron oxide is reduced to Fe_3O_4 which remains insoluble with SiO_2 and Al_2O_3. The manganese nitrate solution is concentrated and decomposed in air at 200 °C to produce MnO_2 and nitric acid which is for reuse.

15.3.2 Metallic Manganese

There are two methods for the production of manganese: (a) alumino-thermic method; and (b) electrolytic method.

15.3.2.1 Alumino-Thermic Method

This is carried out in cylindrical magnesite lined steel pots. Charge consists of about 3500 kg of ore, 1000 kg of aluminum powder and 0.9 kg of sodium peroxide as igniter. A portion of the charge is feed to the pot, ignited and then gradually fed the remaining charge. Cooled for 24 h, the metal (1700 kg) is broken out with hammer. Metal contains 97–98% Mn, 0.5–1.0% Al, 0.75% Fe and 0.28% Si. Manganese yields 75%, since manganese metal is lost due to volatilization and goes to slag.

$$3MnO_2 + 4Al = 3Mn + 2Al_2O_3 \tag{15.1}$$

15.3.2.2 Electrolytic Method

Electrolyte is prepared by leaching the reduced ore with spent electrolyte containing 12 g manganese (as sulphate), 135 g ammonium sulphate and 40 g sulphuric acid per litre. Impurities consisting of iron, arsenic, copper, zinc, lead etc. are precipitated with hydrogen sulphide. The purified leaching solution containing 35–40 g of manganese per litre is pumped to lead-lined cells, sulphur dioxide is added to prevent precipitation of manganese hydroxide. Cathodes are stainless steels and anodes consist of lead with 2% arsenic and silver. Anodes are perforated, all possible measures are taken to reduce the surface area, thus increasing current density and minimizing the formation of precipitated oxides. These precautions regarding the anodes are important, as oxide up to 30% tends to be precipitated if they are omitted.

In the aqueous electrolysis of manganese, it is necessary to control the pH in the anode and the cathode compartments. The pH must be controlled in order to suppress the evolution of hydrogen at the high voltage needed for deposition of manganese on the cathode. After deposition for 24 h, the cathodes are removed and the manganese metal is sprayed to prevent oxidation. The manganese produced by electrolysis is 99.0–99.5% pure; main impurities are hydrogen (0.015%) and sulphur (0.04%). Hydrogen is removed by heating the manganese metal at 500 °C.

15.3.3 Ferro-Manganese

There are two methods for ferro-manganese production: (a) blast furnace; (b) electric furnace.

15.3.3.1 Blast Furnace Method

Ferro-manganese is produced in blast furnace with losses varying up to 20%. The standard grade of ferro-manganese contains 80% Mn, 14% Fe and 6% C.

The actual ways in which the reductant takes part in the reduction reactions are as follows:

(1) Gasification (presumably by the Boudouard reaction: $CO_2 + C = 2CO$) to form CO which can react high in the furnace.

(2) Dissolution of carbon in the metal product.
(3) Reaction of the dissolved carbon with MnO in the slag ($MnO + \underline{C} = Mn + CO$).
(4) As the metal descends to the hearth ('metal refining').
(5) Gasification by the Boudouard reaction is as part of the 'direct reduction' reaction of solid carbon with MnO in the slag ($MnO + C = Mn + CO_2$).

15.3.3.2 Electric Furnace Method

Electric furnace method is used for production of low carbon ferro-managanese. Raw materials for the production of ferro-managanese comprise ores/concentrates containing at least 45% Mn, carbonaceous material (coke breeze), steel scrap (turning) or iron ore and limestone (as flux) are charged. A higher content of manganese in ores/concentrates ensures a higher productivity of furnace and lower consumption of electric energy per tonne of ferro-alloy; the content of silica in ores/concentrates should be as low as possible as silica increases the slag volume and that causes a higher loss of manganese into the slag and a higher use of electric energy. The reductant for the process is usually coke breeze with not more than 12% ash, and less than 0.02% phosphorus; its size should be from 3 to 15 mm. Phosphorus is a harmful impurity in ferro-manganese.

The smelting process includes the stages as follows:

(1) Preheating of the materials.
(2) Drying and removal of volatiles.
(3) Reduction of oxides.
(4) Melting of the metal and slag.

Moisture content of the charge materials is removed within 10–15 min of charging. Volatiles are removed in the temperature range from 200 to 1000 °C. Ferric oxide contained in manganese ore is reduced with carbon monoxide and hydrogen at low temperatures:

$$3Fe_2O_3 + CO/H_2 = 2Fe_3O_4 + CO_2/H_2O \quad (15.2)$$

$$Fe_3O_4 + CO/H_2 = 3FeO + CO_2/H_2O \quad (15.3)$$

Ferrous oxide is first reduced with carbon monoxide and hydrogen at 500–600 °C, and after that with solid carbon in the deeper zones of the bath:

$$FeO + CO/H_2 = Fe + CO_2/H_2O \quad (15.4)$$

$$FeO + C = Fe + CO \quad (15.5)$$

The reduction of manganese from pyrolusite occurs stepwise at low temperature:

$$3MnO_2 + 2CO = Mn_3O_4 + 2CO_2 \quad (15.6)$$

$$Mn_3O_4 + CO = 3MnO + CO_2 \quad (15.7)$$

Manganous oxide is reduced with solid carbon:

$$3MnO + 3C = 3Mn + 3CO \quad (15.8)$$

$$3MnO + 4C = Mn_3C + 3CO \quad (15.9)$$

The theoretical temperatures for these reactions (15.8) and (15.9) are in the range of 1223–1420 °C. At higher temperature, manganese carbide will be favoured.

The manganous oxide goes to slag at higher temperature:

$$MnO + SiO_2 = MnO.\,SiO_2 \quad (15.10)$$

Iron turnings are added to reduce the manganous oxide from the slag:

$$MnO.\,SiO_2 + Fe = Mn + FeO.\,SiO_2 \quad (15.11)$$

The reducing conditions in the furnace ensure that phosphorus is reduced almost fully. The acid slag cannot absorb phosphorus, so that around 20–25% phosphorus is removed with furnace gases and 75–80% passes to the alloy.

The recovery using an electric furnace may be as high as 90%, and coke consumption is half that of the blast furnace. Operation is continuous with intermittent tapping. Metal losses are considerably reduced in electric furnace.

A silico-manganese alloy, combining both the advantages of ferro-manganese and ferro-silicon, is used for deoxidation and imparting favourable mechanical properties, especially to high-speed steels. It is produced in electric furnace from silica, manganese ore, steel scrap and coke. A silico-manganese alloy contains 65–70% Mn, 15–25% Si. This alloy can be used instead of carbon as a reducing agent in the production of low carbon ferro-manganese and also low carbon manganese steel.

15.4 Properties

Pure metallic manganese is silvery white in colour but the presence of carbon renders it grayish. Manganese has an atomic mass of 54.94, specific gravity of 7.3. Melting and boiling points of manganese are 1244 °C and 2150 °C respectively. Manganese can be rolled, and it is used to a limited extent as sheet and wire in electric lamps and valves.

Manganese possesses a strong affinity to oxygen and sulphur. Iron and manganese are soluble in all proportions in solid as well as in liquid state. Manganese combines with carbon to give stable carbide (Mn_3C). Manganese carbide and liquid manganese are soluble in all proportions.

Together with silicon, manganese forms several silicides: Mn_2Si, MnSi, Mn_2Si_3; the most stable among these is MnSi. Silicides of manganese are stabler compounds than manganese carbide. Silicon, therefore, displaces carbon in its manganese compounds and forms manganese silicide. The higher the silicon contents in the alloy, the lower the carbon content.

Manganese reacts with oxygen to form oxides: MnO_2, Mn_2O_3, Mn_3O_4 and MnO. On heating above 530 °C, MnO_2 transforms into Mn_2O_3, and on heating above 940 °C into Mn_3O_4. Reactions of manganese with nitrogen give nitrides: Mn_4N_2, Mn_3N_2 and Mn_4N.

Hardenability, hardness, ultimate strength and yield point of steel increase and its toughness decreases with a higher content of manganese. High abrasion resistance steel contains 11–14% Mn.

15.5 Applications

Manganese is probably the most important of all the auxiliary metals used in steelmaking for deoxidation and desulphurization purposes. Metallic manganese is used in the production of special steels. Manganese steel (1–14% Mn) has a high tensile strength combined with great elongation (50–70%); it is used in crusher jaws. It owes its power of resisting heavy abrasive wear to its properties of hardening under cold working. In a jaw crusher where the wearing surfaces are subjected to heavy pressure and abrasion, manganese steel will outlast metals whose initial hardness is much greater.

Manganese also plays an important role in copper base alloys. Manganin, a copper alloy containing 15% Mn, and up to 3% Ni, possesses high electrical resistivity coupled with a low temperature coefficient making it suitable for incorporation in instruments and appliances where constant electrical resistivity is important. Where strength and toughness combined with resistance to corrosion (specially sea water corrosion) are needed, manganese-bronze is frequently used. Containing no tin, manganese-bronze is not really a bronze but a high-strength brass. Very large castings are made specially for such purposes as ships' propellers and rudders; smaller castings including mountings for ships' compasses, gear wheels, pinions, and similar applications.

Ferro-manganese is widely used in the manufacture of tool steels (up to 0.4% Mn) and structural steels (up to 0.6% Mn), and also of special steels with a high manganese content (up to 12–14% Mn), as well as an alloying addition. Low carbon ferro-managanese is preferred for production of low carbon steel. Ferro-manganese and silico-manganese are widely used as reducers in the manufacture of steel. Ferro-manganese is also employed for deoxidation and desulphurization of steel, while silico-manganese is used as a semi-product in the manufacture of medium and low carbon ferro-manganese and metallic manganese.

Chapter 16
Chromium

16.1 Introduction

Chromium (Cr) is an element of the 6th Group of the Periodic Table (Fig. 1.1). Mean chromium content in the Earth's crust comes to 0.02%. The element chromium was first isolated in 1797 by Vauquelin from the mineral crocoites, which was discovered in 1765 in Russia by Pallas. Its name was derived from the Greek word *chromos*, meaning colour, because of the varied colours of its compounds.

Pure chromium was obtained for the first time in 1854 by the electrolysis of aqueous solutions of chrome chloride. In 1908 a metal containing 99.9% Cr was obtained by the reduction of chrome oxide by hydrogen gas.

16.2 Sources

Chromium occurs in nature as chromite ($FeO.Cr_2O_3$). The natural mineral is rarely pure, Cr_2O_3 being replaced by Al_2O_3, Fe_2O_3; the higher grade of ore contains about 55% Cr. Many ores are sufficiently up to the grade to market without any concentration processes.

Chrome ores consist of chromospinelides: (Mg, Fe) Cr_2O_4 (magnesiochromite), Fe $(CrAl)_2O_4$ (aluminochromite), etc. Chromospinelides may contain up to 62% Cr_2O_3, the balance usually being iron, aluminium and magnesium oxides.

S.K. Dutta and D.R. Lodhari, *Extraction of Nuclear and Non-ferrous Metals*,
Topics in Mining, Metallurgy and Materials Engineering,
https://doi.org/10.1007/978-981-10-5172-2_16

16.3 Extraction

16.3.1 Metallic Chromium

(a) The high grade chromium ore is grounded and then roasted with soda ash (Na_2CO_3) at 800–1000°C in rotary kilns.

$$2Na_2CO_3 + Cr_2O_3 + 1.5O_2 = 2Na_2CrO_4 + 2CO_2 \tag{16.1}$$

Limestone is added during roasting to avoid sintering. The roasted product is leached with water and filtered to separate the soluble Na_2CrO_4 from the iron oxide and gangue present in the ore. This Na_2CrO_4 is converted to $Na_2Cr_2O_7$ by sulphuric acid treatment.

$$2Na_2CrO_4 + H_2SO_4 = Na_2Cr_2O_7 + Na_2SO_4 + H_2O \tag{16.2}$$

$Na_2Cr_2O_7$ in the aquous solution, is selectively crystallized by evaporation, and the crystals are separated by filtration. The dried $Na_2Cr_2O_7$ crystals are then converted to Cr_2O_3 by reduction with the help of sulphur.

$$Na_2Cr_2O_7 + S = Cr_2O_3 + Na_2SO_4 \tag{16.3}$$

Cr_2O_3 is finally reduced by Al (i.e. aluminothermic reduction) to yield metallic chromium.

$$Cr_2O_3 + 2Al = 2Cr + Al_2O_3 \tag{16.4}$$

(b) The carbothermic reduction of chromium seems to be of great technical interest. Figure 16.1 shows the thermodynamic equilibrium diagram according to Pourbaix-Ellingham; and for the Cr–O–C system.

The following reduction reactions occur as per the above diagrams:

$$3Cr_2O_3(s) + 13C(s) = 2Cr_3C_2(s) + 9CO(g) \tag{16.5}$$

$$5Cr_2O_3(s) + 27Cr_3C_2(s) = 13Cr_7C_3(s) + 15CO(g) \tag{16.6}$$

$$Cr_2O_3(s) + 3Cr_7C_3(s) = Cr_{23}C_6(s) + 3CO(g) \tag{16.7}$$

$$2Cr_2O_3(s) + Cr_{23}C_6(s) = 27Cr(s) + 6CO\ (g) \tag{16.8}$$

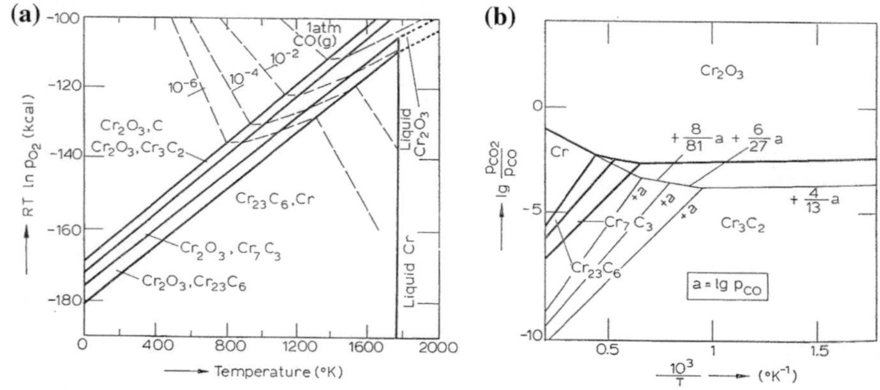

Fig. 16.1 **a** Pourbaix-Ellingham diagram for the system Cr–O–C. **b** Thermodynamic equilibrium diagram for the system Cr–O–C for $p_{CO} = 1$ atm (*heavy line*) and $p_{CO} = 10^{-4}$ atm (*thin line*). *Source* Vacuum Metallurgy, Eds O. Winkler and R. Bakish, Elsevier Pub Co, p. 176

Table 16.1 Reduction temperatures of metal oxides of the 6th Group of the Periodic Table

CO Pressure (atm)	Temperature (°C)		
	Chromium oxide	Molybdenum oxide	Tungsten oxide
1	2000	1070	900
10^{-2}	1550	825	685
10^{-4}	1260	655	540
10^{-6}	1050	535	430

From these diagrams it is evident that reduction to the metal oxide is not possible at atmospheric pressure. Table 16.1 shows the minimum temperatures for the reduction of chromium, molybdenum and tungsten oxides. According to these temperatures, favourable conditions for a reduction of all three metal oxides in a vacuum should be expected.

16.3.2 Ferro-Chromium

Ferro-chromium (with 60% Cr and 6% C) was produced for the first time in 1893 by electric furnace. All grades of the ferro-chrome can be produced in the electric furnace. Many grades of the ferro-alloy are available with carbon content ranging from 0.06% to as high as 8%. The decisive factors for estimation of the quality of chrome ores are as follows:

(1) The content of chromic oxide; in order to produce ferro-chrome, the content of chromic oxide in ore should be at least 45%.

(2) The ratio of concentration of chromic oxide to that of ferrous oxide; it should be not less than 2.5 to ensure that the alloy will contain 60% Cr.
(3) The proportion and composition of the gangue, which determines the amount of slag to be formed in smelting, should be under controlled; a large bulk of slag will involve much energy spent on its melting and overheating, also a greater loss of chromium to slag.

16.3.2.1 High Carbon Ferro-Chrome

While there is some reduction of iron in the solid state (with CO as the reductant), the reduction rate apparently only increases strongly once the chromite starts dissolving in the slag above 1500°C; the reduction reaction then occurs through reaction of dissolved chromium oxide in the slag with solid carbon, or with carbon and silicon dissolved in the metal.

The relevant reactions of the reductant can be summarized as follows:

(1) Dissolution of carbon in the metal product.
(2) Reduction of silicon (presumably by production of SiO gas in the higher-temperature region, with reduction of SiO to silicon or silicon carbide higher in the furnace).
(3) Reaction of dissolved carbon with Cr_2O_3 in the slag as the metal descends to the hearth.
(4) Gasification by the Boudouard reaction: $CO_2 + C = 2CO$ (during direct reduction).

For the high carbon ferro-chrome production, finely ground ore is mixed with carbon (coke breeze) in the proportion of 1:3 at 1230°C:

$$FeO.Cr_2O_3 + 4C = 2Cr + Fe + 4CO \tag{16.9}$$

Stable carbides are also formed in the reduction of chromium oxide by carbon at 1100°C:

$$Cr_2O_3 + 27/7\,C = 2/7\,Cr_7C_3 + 3CO \tag{16.10}$$

FeO, present in ore, is also reduced by carbon:

$$FeO + C = Fe + CO \tag{16.11}$$

Iron reduced from ore promotes the reduction of chromium oxide, since chromium carbides, which appear in the later reaction (16.10), dissolve in iron with the formation of double carbides [$(Cr, Fe)_7C_3$].

$$Cr_2O_3 + 2\underline{Fe} + 27/7\,C = 2/7\left[(Cr, Fe)_7C_3\right] + 3CO \tag{16.12}$$

At a sufficiently high temperature, the melt being formed can be decarbonized by the reaction:

$$1/3\,(Cr, Fe)_7C_3 + 1/3\,Cr_2O_3 = 3\,Cr + 7/3\,Fe + CO \tag{16.13}$$

Sulphur can be brought into the melt mainly by coke breeze. In operation with acid slag the main portion of sulphur passes to the melt and part of it is volatilized. The content of sulphur in high carbon ferro-chrome must not exceed 0.06%, and therefore the tolerable content in coke breeze is not more than 0.5% S.

16.3.2.2 Low Carbon Ferro-Chrome

Ferro-chrome with less than 2% C is produced by the silico-thermic process. For the production of the low carbon ferro-chrome, silicon is used as a reducer. First ferro-silico-chrome is made and then its silicon reduces chromium and iron oxides from chrome ore, the reaction is exothermic, i.e. can proceed at lower temperature.

$$FeO.Cr_2O_3 + 2\,Si = 2\,Cr + Fe + 2\,SiO_2 \tag{16.14}$$

16.3.2.3 Extra-Low Carbon Ferro-Chrome

In the manufacture of extra-low carbon ferro-chrome (0.05–0.06% C), chrome ore is reduced with the silicon of ferro-silico-chrome. The reaction (16.14) is exothermic, i.e. can proceed at lower temperature. At lower temperature the thick high chromium slag is made, which retards the diffusion processes and reduces the reaction rates sharply because ferro-chrome is smelted at a high temperature. Since iron lowers the melting point of the alloy and simultaneously lowers the concentration of chromium in it.

Most of the silicon in ferro-silico-chrome is present as chromium and iron silicides; a certain amount of energy must therefore be spent to decompose them. By addition of lime, silica is bonded by lime to form $2CaO.SiO_2$, i.e. the activity of silica will be lowered. A higher basicity of slag (i.e. a lower content of free silica in it) results in less silicon remains in the metal and less chromic oxide in the slag. The optimum CaO/SiO_2 ratio is 1.7 to 1.8. An excessive high basicity results in higher losses of chromium into the slag.

16.4 Properties

Pure chromium is a grey lustre metal. Chromium has an atomic mass of 51.996, specific gravity 7.19, melting point 1857°C, and boiling point 2672°C. Metallic chromium is bluish-white in colour, hard, brittle, and possesses a high lustre. It is highly resistant to corrosion by atmosphere, sea water and sulphur compounds.

Chromium reacts with the anhydrous halogens, hydrogen chloride, and hydrogen fluoride. Aqueous hydrofluoric acid, hydrochloric acid, hydrogen bromide, and hydrogen iodide will dissolve chromium, as will dilute sulphuric acid. Hydrogen gas is evolved during the dissolution of chromium in dilute sulphuric acid, whereas in boiling concentrated sulphuric acid, sulphur dioxide is liberated.

Iron and chromium can form a series of solid solutions. Chromium and carbon can form carbides: Cr_4C (5.46% C), Cr_7C_3 (9.01% C) and Cr_3C_2 (13.34% C), having melting points at 1520°, 1780° and 1895°C respectively. Chromium and silicon form silicides Cr_3Si (melting point of 1710°C), CrSi (melting point of 1545°C), $CrSi_2$ (melting point of 1550°C), etc.; and the most stable silicide CrSi; these are more stable compounds than chromium carbides. Chromium can form CrN and Cr_2N nitrides with nitrogen. Chromium and oxygen form oxides: CrO_3 (acid oxide), Cr_2O_3 (amphoteric oxide), and CrO (basic oxide); among them Cr_2O_3 being the most important for metallurgy. Chromic oxide is a bright green powder, and is the stablest of the oxides with a specific gravity of 5.21 and melting point 2275°C.

16.5 Applications

Metallic chromium is chiefly used for alloying purposes. Chromium is used as an alloying element, in the form of ferro-alloy, in steel; it contributes hardness and toughness, and resistance to wear, heat and corrosion. Chromium is used up to 2% in medium and high carbon steels, usually for the hardening and toughening effects and for increased fatigue and wear resistance. Chromium steels containing 1% C and 1% Cr are very hard and are widely used in the manufacture of ball and roller bearings.

Chromium is used extensively as an alloying element in many grades of steels and alloys. Addition of chromium makes steel more liable to self-hardening and increases hardening depth. Chromium can also increase the strength and hardness of steel and lower the ductility. In combination with silicon, chromium imparts heat-resisting properties to steel. High chromium steels (over 12% Cr) with a low carbon content become rust-resistant after hardening. With high content of carbon in steel, chromium can increase the wear resistance; it also increasing the ultimate strength at high temperatures.

The combination of chromium with nickel extends into the field of the non-ferrous alloys, to high resistance to oxidation at elevated temperatures. Ni-Cr alloys exhibit a remarkable resistance to both high temperature oxidation and corrosion, for use as heating elements in industrial and household electrical appliances. The property of red-hardness gives to the Cr–Co–W-alloys a prominent position in the list of high-speed cutting tools.

High carbon ferro-chrome is used as alloying element for the manufacture of alloy steel and chromium steel. Low carbon ferro-chrome is used as a chromium addition for the manufacture of stainless steel, refractory steel and steels for cryogenic application. Ferro-silicon-chrome has also found application for slag deoxidizing in the manufacture of stainless steel. Ferro-silicon-chrome (containing 48–51% Si, 30% Cr, 0.04% C) is used for making medium, low and extra-low carbon ferro-chrome.

Chapter 17
Tungsten

17.1 Introduction

Tungsten (W), also known by its European name wolfram, is a metal with unique properties that lead to its use in cutting and forming other metals and in important high temperature applications. The word *tungsten* is an adaptation of the Swedish *tung sten* (that means heavy stone) and was first applied to the mineral scheelite about 1758. The element was first identified in 1781 by a Swedish chemist, K.W. Scheele, for whom the calcium tungsten mineral, scheelite was later named. The iron-manganese-tungsten mineral, wolframite, was first described about 1574. The overall tungsten content in the Earth's crust is negligible (0.0009%).

The element did not find industrial application until early in the nineteenth century, but it has since become one of the most important industrial metals. China is the chief producer of tungsten, followed by Russia, Portugal, Burma, USA and Bolivia.

17.2 Sources

Main sources of tungsten ores are wolframite [$(FeMn)WO_4$] and scheelite ($CaWO_4$), which contain 0.3–2.0% W. Other tungsten minerals are ferberite ($FeWO_4$) and hubnerite ($MnWO_4$). The most common of the tungsten minerals is wolframite, output of which amounts to more than 90% of the world's output. The biggest tungsten ore deposits are found in China, Australia, Korea, Russia and Burma.

Wolframite is an iron manganese tungstate, carrying more iron than manganese; and scheelite, a calcium salt. Since these ores have high specific gravities (wolframite 7.2–7.5 and scheelite 5.9–6.1) compared with the specific gravity of the gangue (1.8–2.5), they are concentrated by gravity separation method, except when cassiterite (specific gravity 6.8–7.1) is present. When cassiterite is present along with scheelite, a multistage procedure is used to upgrade the ore.

S.K. Dutta and D.R. Lodhari, *Extraction of Nuclear and Non-ferrous Metals*,
Topics in Mining, Metallurgy and Materials Engineering,
https://doi.org/10.1007/978-981-10-5172-2_17

17.3 Extraction

17.3.1 Metallic Tungsten

Tungsten is used in steelmaking in either the metallic state or as the ferro-alloy form. The tungsten content of the ore as mined is about 0.3–2.0%, with concentration depending chiefly on gravity separation methods due to high density of the materials. Since wolframite is magnetic, that can be concentrated by magnetic separation. If tungsten occurs as scheelite (which is non-magnetic) in a wolframite concentrate; magnetic separation results in a magnetic wolframite and a non-magnetic tin-scheelite concentrate. The tungsten in the latter has to be separated by chemical means, i.e. roasting with soda ash at 1000°C to form sodium salt (Na_2WO_4), then leaching with boiling water and filtering to remove impurities such as silica, iron, calcium etc. Hydrochloric acid is added to the clarified solution, tungsten trioxide [WO_3 (60%)] is precipitated which is filtered, washed and dried.

Tungsten trioxide is reduced by either (a) carbon or (b) hydrogen.

(a) The carbothermic reduction of tungsten trioxide (WO_3) proceeds to completion at 1100°C and at reduced pressure (Table 16.1). The tungsten trioxide is mixed with charcoal and heated, the resulting grey product (tungsten metal) being ground and screened for suitable use.

$$WO_3 + 3\,C = W + 3\,CO \tag{17.1}$$

(b) The tungsten trioxide is reduced by stream of hydrogen at a temperature of 700–800°C in stage wise:

$$3\,WO_3 + H_2 = W_3O_8 + H_2O \tag{17.2}$$

$$W_3O_8 + 2\,H_2 = 3\,WO_2 + 2\,H_2O \tag{17.3}$$

$$3\,WO_2 + 6\,H_2 = 3\,W + 6\,H_2O \tag{17.4}$$

Overall reaction:

$$WO_3 + 3\,H_2 = W + 3\,H_2O \tag{17.5}$$

The tungsten powder produced from above reaction is suitable for producing the electric bulb filament and tungsten carbide.

17.3.2 Ferro Tungsten

For production of ferro-alloy, wolframite [(FeMn)WO_4] containing iron offers a convenient source of the raw material. The pure wolframite contains 75% WO_3 and 25% MnO and FeO. The reduction is carried out in an electric furnace in the presence of either carbon or silicon.

For the production of ferro-tungsten, raw materials are used as follows:

(1) Tungsten concentrate (100 kg).
(2) Coke (8–10 kg), used for the reduction of tungsten and iron oxides. Coke should be minimum content of phosphorus, sulphur and ash.
(3) 75% ferro-silicon (7 to 10 kg) is used for rapid and complete reduction of tungsten oxide contained in the slag prior to the tapping,
(4) Iron turning introduced into the mixture to reduce the alloy's viscosity.

The most widely used method of ferro-tungsten production is the simultaneous reduction of tungsten and iron oxides by carbon:

$$WO_3 + 3\,C = W + 3\,CO \tag{17.1}$$

The process begins at 690°C. Iron oxide is reduced at 950°C:

$$FeO + C = Fe + CO \tag{17.6}$$

The most common grades contain 78–83% W, 0.75% Si and 0.1% C and usually supplied in lumps form.

17.4 Properties

Atomic weight of tungsten is 183.92 and specific gravity is 19.32. Tungsten has a very high melting point (3410°C) and high modulus of elasticity, which are the highest of all metallic elements; and its low vapour pressure and low compressibility are the lowest of all metallic elements. Tungsten has a high boiling point (5927°C).

The thermal conductivity of tungsten, although less than half that of copper, is much higher than that of iron or nickel. The electrical conductivity of tungsten, although one-third that of annealed copper, is higher than that of iron, nickel, mercury, platinum, or phosphor bronze.

Tungsten exhibits high strength and high hardness at both room temperature and elevated temperatures. Fine tungsten wire has the highest tensile strength. In spite of its high specific gravity, the strength–weight ratio of tungsten at elevated temperature is superior to that of all other metals.

Tungsten oxides are formed when heated in the air from 400–500°C. The melting point of tungsten alloys increase with the rise in oxide content. The melting point of tungsten alloy containing 70–80% W is 2000°C.

With carbon tungsten forms tungsten carbides: WC contains 6.12% C; and W_2C contains 3.16% C; its melting point is 2750°C. Carbide inclusions make tungsten brittle. Tungsten forms two compounds with silicon: W_2Si_3 and WSi.

Tungsten forms continuous series of binary solid solutions only with molybdenum, columbium, and tantalum, and with chromium above 1400°C. Ternary systems of tungsten with molybdenum, columbium, and tantalum are also continuous series of solid solutions. Additions of molybdenum, columbium, and tantalum, decrease the hardness and tensile strength of tungsten at room temperature; but they improve the high temperature strength of tungsten.

17.5 Application

Tungsten is universally use as electric bulb filaments. It is used in the form of ferro-alloy as alloying element for the production of tool and die steels. Alloyed with molybdenum, tungsten is used in thermionic valves and in special thermocouples. W–Ni–Co–C alloys are highly resistant to corrosion and are used as valve facings and valve seats and for tools, such as the cutting edges of hot shears. It is used as special temperature resisting alloys for jet propulsion and gas turbines.

Tungsten is one of the most important alloying elements in the production of special, tool and certain grades of steels. Hardness, tensile strength and elasticity of steels are increased by addition of tungsten. Steels with high tungsten content (3.5–19%) are used for the manufacture of high-speed cutting tools. Tungsten is present in steel in the form of carbide. With low tungsten content, tungsten carbide (WC) is present with iron carbide (Fe_3C) in varying proportions.

Chapter 18
Molybdenum

18.1 Introduction

Molybdenum (Mo) is in Group VIB of the Periodic Table (Fig. 1.1) and therefore related to chromium and tungsten. The terms *molybdena, plumbago, graphite*, and *galena* seem to have been applied in early Greek writings to lead sulphide and minerals of similar appearance, including graphite, as well as the naturally occurring molybdenum disulphide. This confusion, especially between the minerals now called graphite and molybdenite, continued until 1778 when K.W. Scheele demonstrated that molybdenite contains sulphur and prepared molybdic acid from the mineral. In 1782, P.J. Hjelm obtained an impure metallic product by reducing the oxide with carbon and called it molybdenum.

18.2 Sources

Molybdenum does not occur free in nature. Molybdenite (MoS_2) and, to a much lesser extent, welfenite ($PbMoO_4$) are the only minerals of commercial value. Molybdenite is a soft, shiny, blue mineral, occurring in flakes closely resembling graphite, and up to the end of the eighteenth century the two were thought to be one and the same compound. They can be differentiated by their streak (that of graphite is grey, molybdenite's is greenish grey) and also by the way the latter gives off sulphur dioxide when heated.

Molybdenite occurs in disseminated porphyry deposits. They are characterized by their low grade and large volume. In these deposits, the molybdenum content varies from 0.1–0.5%, and the ore is associated with pyrite and small amounts of tungsten, tin, topaz, and uranium. Molybdenum also occurs in some copper porphyry deposits which contain 0.005–0.1% Mo. Copper porphyry deposits contribute the lion shear of the world's molybdenum production. Molybdenum is concentrated in the USA and Chile.

S.K. Dutta and D.R. Lodhari, *Extraction of Nuclear and Non-ferrous Metals*,
Topics in Mining, Metallurgy and Materials Engineering,
https://doi.org/10.1007/978-981-10-5172-2_18

18.3 Extraction

18.3.1 Concentration of Molybdenite

Molybdenite in the ore is concentrated by flotation, because it possesses excellent flotation characteristics (contact angle is 145°, compared with 135° for pyrite and 131° for chalcopyrite). Hence, MoS_2 has a higher flotation capability than other common sulphide minerals. Molybdenite itself is more readily floated than most minerals, with two reagents being required: (a) pine oil as frother; and (b) kerosene as collector.

Two main factors influence the flotation as follows:

(1) *Fine grinding:* molybdenite is closely associated with the quartz and contaminating minerals. A fine grind (−270 mesh) is necessary to secure satisfactory liberation of the valuable mineral.
(2) *Floatability of molybdenite:* Under certain conditions, a middling (65–100 mesh), containing less than 1% molybdenite, can be floated. In early practice, ore was ground through 100 mesh for rough flotation, pine oil was only used for reagent. Recovery was satisfactory, but maintaining the grade of concentrate was difficult. The present practice is to make rough concentrate, grind that and refloat.

From molybdenum porphyry ores, over 90% of the molybdenum can be recovered by comminution and flotation; and from copper porphyry ores, the recovery of MoS_2 is lower, i.e. 30–60% due to the lower MoS_2 content in the ore. The general flow diagram for the concentrate of molybdenite from copper porphyry ores is shown in Fig. 18.1.

Both Cu and Mo are readily recovered by flotation with reagents such as xanthate. Under certain conditions, a middling, containing less than 1% molybdenite, can be floated. In the next stage, the xanthate is eliminated by oxidation to render copper non-floatable, and molybdenum is floated with the addition of hydrocarbon. Copper is further eliminated from the MoS_2 concentrate by leaching with ferric chloride.

The MoS_2 concentrate is roasted to MoO_3 to provide the raw material for production of metallic molybdenum, calcium molybdate salt containing 40–45% Mo, and ferro-molybdenum (60–70% Mo).

$$2MoS_2 + 7O_2 = 2MoO_3 + 4SO_2 \tag{18.1}$$

The roasting of MoS_2 to MoO_3 is chemically simple. It requires good temperature control since the rate of roasting is slow at temperature below 600°C and the calcine oxide starts to sinter, leading to molybdenum loss due to volatilization of the oxide at temperature above 650°C. The desired temperature control is achieved

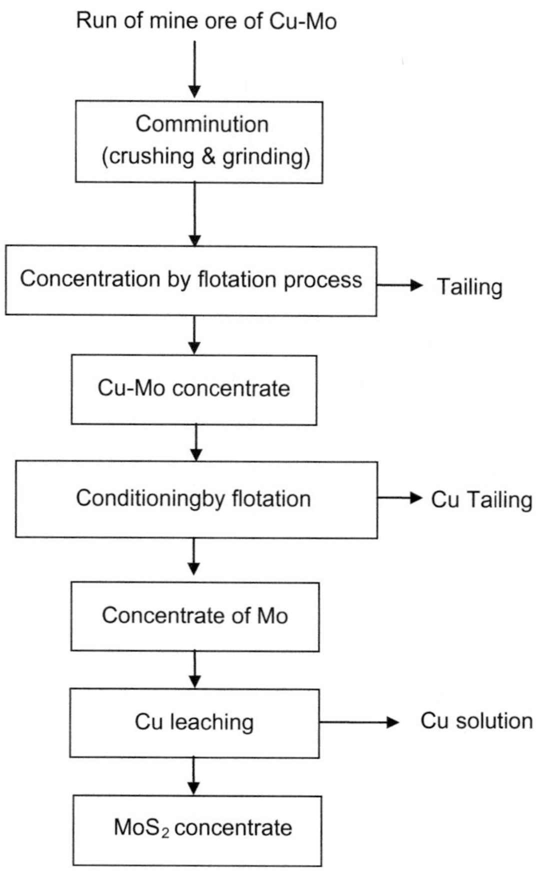

Fig. 18.1 Recovery of molybdenum from Cu-Mo ore

by employing a multiple hearth roaster. The desired sulphur content in the oxide product (0.1–0.5%) is achieved by good temperature control. The roasted product contains 80–90% MoO_3, 3–8% FeO, 3–10% SiO_2, 0.4–2.0% CuO, 0.05–0.15% S and 0.02–0.15% P. Briquettes are made from the roasted product by adding pitch.

The roasted concentrate is divided into three grades, depending on its sulphur content: (a) grade 1 up to 0.09% S; (b) grade 2 up to 0.10–0.12% S; and (c) grade 3 up to 0.13–0.15% S. If the sulphur content is more than 0.15%, the concentrate is discarded and re-roasted.

18.3.2 Metallic Molybdenum

Molybdenum metal is produced by the hydrogen reduction of the briquettes of MoO_3 at 1000°C.

$$MoO_3 + 3H_2 = Mo + 3H_2O \tag{18.2}$$

The reduced grey metal powder is pressed into bars, sintered. The metalized briquettes are then pressed into consumable electrodes for vacuum electric arc melting into high purity molybdenum ingots.

18.3.3 Calcium Molybdate

Calcium molybdate is produced by adding hydrated lime to one of the lower hearths during the roasting of the concentrate; the molybdenum trioxide reacts with the lime to form calcium molybdate.

$$MoO_3 + Ca(OH)_2 + 4\,C = Mo + Ca + H_2O + 4\,CO \tag{18.3}$$

18.3.4 Ferro-Molybdenum

(a) Ferro-molybdenum is produced either in the electric furnace or by the thermit process. In the electric furnace, the oxide is reduced with either carbon or silicon. Steel scrap or pyrite is added to produce ferro-molybdenum of the desired composition.

$$MoO_3 + 3\,C = Mo + 3\,CO \tag{18.4}$$

$$2\,MoO_3 + 3\,Si = 2\,Mo + 3\,SiO_2 \tag{18.5}$$

(b) If the MoS_2 concentrate is used, lime is added for sulphur removed.

$$MoS_2 + CaO + 2\,C = Mo + CaS + CO + CS \tag{18.6}$$

(c) In the thermit process, the oxide is mixed with powder aluminum, silicon, iron ore, lime and fluorspar; then it is fired in a cylindrical brick-lined steel shell, set over a sandpit. Iron ore is used as an iron-bearing material and additional source of oxygen. At the same time, iron ore is a fluxing material because the iron oxide partly goes to slag and increase the fluidity of the slag (since highly siliceous slag forms which has high viscosity).

Smelting is completed in 15 min; when the slag is poured off, ferro-alloy is left in the sand to solidify. The massive ferro-alloy is then reduced in size by jaw crusher. A typical assay is: 74% Mo, 0.06% C, 0.08% S and 1.0% Si.

18.4 Properties

The atomic weight of molybdenum is 95.95 and its specific gravity is 10.22; melting and boiling points are 2610°C and 5560°C respectively. Its high melting point is probably of most importance in its commercial uses. Only tungsten and tantalum, of the more readily available metals, have higher melting points.

Molybdenum, in common with many other metals, and undergoes a change from ductile to brittle behaviour as the temperature of deformation is lowered. Much of the interest in molybdenum and its alloys is due to their outstanding strength at high temperatures. Molybdenum base alloys have been developed with much better high temperature strength than unalloyed molybdenum.

The high elastic modulus of molybdenum may become of increasing importance in the aircraft and missile industries. A comparison of the modulus-density ratio for several commercial alloys shows molybdenum to be outstanding over a wide temperature range.

Other applications of molybdenum may take advantage of its thermal conductivity, which is several times as great as that of most high temperature superalloys, or of the electrical conductivity, which is approximately one-third that of copper.

18.5 Applications

Principal use of molybdenum is as a ferro-alloy in the iron and steel industries. Ferro-molybdenum is an alloying addition which sharply raises the mechanical and physical properties of steels. The addition of molybdenum ensures uniform micro-crystalline structure in steel, increase its hardenability and decrease in post-temper brittleness. Structural steels with 0.15–0.50% Mo possess greater impact strength. In steel, Mo is generally used in conjunction with one or more alloying elements, such as Ni, Cr, V or Mn, when it has the effect of intensifying the beneficial properties of the element concerned. Molybdenum is used for the manufacture of high-speed steels, which are used for various cutting instruments like drills, milling cutters, cutting tools, etc. capable of withstanding high cutting speeds.

Its alloys find application in diesel engines, automobiles, lorries and aircraft. Cast Cr–Mo liners generally have a life 50–100% longer than white iron liners and are less liable to breakage. Mo is also used in some of the heat-resisting steels for gas turbine components. The pure metal is used in the form of wire for filament supports in electric lamps, electrodes of mercury vapour lamps and winding electric resistance furnaces.

Molybdenum has become an important material in the missile industry for several high temperature structural parts. The atomic energy industry is one of the most active to use molybdenum for its requirements. Molybdenum has been fabricated into heat exchangers, piping, and structural parts.

Chapter 19
Vanadium

19.1 Introduction

Vanadium (V) was discovered in the year 1801 by Manuel del Rio, Mexico City, in lead ore, and was named *erythronium* by him. In 1805 a French chemist, Collet-Descostils, declared that erythronium was nothing but impure chromium. In 1830, Sefstrom found what he thought was an unrecognized metal in the iron ores of Taberg, Sweden, and named it *vanadium*, in honour of the Scandinavian goddess Vanadis, because of its beautiful multicoloured compounds. Roscoe's silver-white powder was first produced in 1967 by hydrogen reduction of vanadium chloride (VCl_2). This was probably the first nearly pure vanadium metal.

Vanadium is a chemical element of the 5th Group of the Periodic Table (Fig. 1.1). Vanadium content in the Earth's crust amounts to 0.015%, and it is considered to be a rare element as it is very much dispersed in nature.

19.2 Sources

Although widely distributed in nature, commercially worthwhile deposits are rare, the majority of supplies being derived as a by-product from the treatment of uranium, lead and zinc ores. One notable exception is the sulphide ore of patronite ($V_2S_5 + nS$), occurring in Peru, where it has formed the largest known vanadium deposit. It is a greenish-black, amorphous mineral, usually carrying some iron, nickel, molybdenum, phosphorus, and carbon

Bravoite [$(Fe,Ni)S_2$], cobalt or vanadium substitutes for nickel and iron. As found in small quantities in Peru, it contains 25% Fe, 15% Ni, 5% V and 45% S. Sulvanite ($3Cu_2S.V_2S_5$), found only in small quantities in South Australia and USA, contains 51% Cu, 14% V, and 35% S. Roscoelite [$2K_2O.2Al_2O_3$ (Mg,Fe) $0.3V_2O_5.10SiO_2.4H_2O$] is a vanadium-bearing mica found as a vein mineral in a

S.K. Dutta and D.R. Lodhari, *Extraction of Nuclear and Non-ferrous Metals*,
Topics in Mining, Metallurgy and Materials Engineering,
https://doi.org/10.1007/978-981-10-5172-2_19

number of rich gold-bearing veins; it occurs in important quantity as a secondary mineral in the sand of USA. It contains 20% V_2O_5, 7.5% K_2O, 14% Al_2O_3, 2% MgO, 1.5% FeO, and 47.5% SiO_2.

Vanadium is also a component of iron-titan-magnetite ore (34–38% Fe_2O_3, 28–31% FeO, 13–14% TiO_2, 0.63–0.68% V_2O_3, 4–10% SiO_2, 4–8% Al_2O_3 etc.).

19.3 Extraction

19.3.1 Recovery of Vanadium Pentoxide

(a) Figure 19.1 shows the flowsheet for the recovery of vanadium pentoxide from carnotite uranium ore (containing 2% V_2O_5, 0.2–0.4% U_3O_8). The process of extraction consists of crushing, roasting with salt, agitating with alkali to effect solution as sodium vanadate and precipitation as the pentoxide by sulphuric acid. Careful control over roasting is essential to ensure a high percentage of vanadium recovery. The red precipitate of vanadium pentoxide, obtained by treatment with acid after drying, contains 85–90% V_2O_5 and largely balances Na_2O. In the presence of lime, it tends to form an insoluble calcium vanadate during roasting instead of the more soluble sodium vanadate. This problem can be overcome by addition of iron pyrite, which forms insoluble calcium ferrite. Close control of roasting temperature is also an important factor for good recoveries.

(b) The lead-zinc ore contains 0.7–1.0% V_2O_5, rendering its extraction of economic importance. Zinc in the concentrate is recovered by leaching, followed by electrolysis; vanadium is present in the zinc solution. The purified leach liquor containing the zinc as sulphate, after freeing from silica, is adjusted to an acidity of 4 g/l and heated with steam for 15–20 h. Hydrolysis of vanadium occurs; the vanadium hydrate is filtered off, dried and fused to the pentoxide.

19.3.2 Metallic Vanadium

(a) *Alumino-thermic reduction:* Vanadium metal is produced by the alumino-thermic reduction of the V_2O_5 concentrate in the presence of $CaCl_2$ in an argon atmosphere at 750 °C. The vanadium metal is a bright silvery white, very ductile in nature, with a specific gravity of 6.0 and a melting point of 1700 °C. The presence of more than 0.5% of oxygen, makes the metal brittle and unfit for either hot or cold working.

$$3V_2O_5 + 10Al = 6V + 5Al_2O_3 \quad (19.1)$$

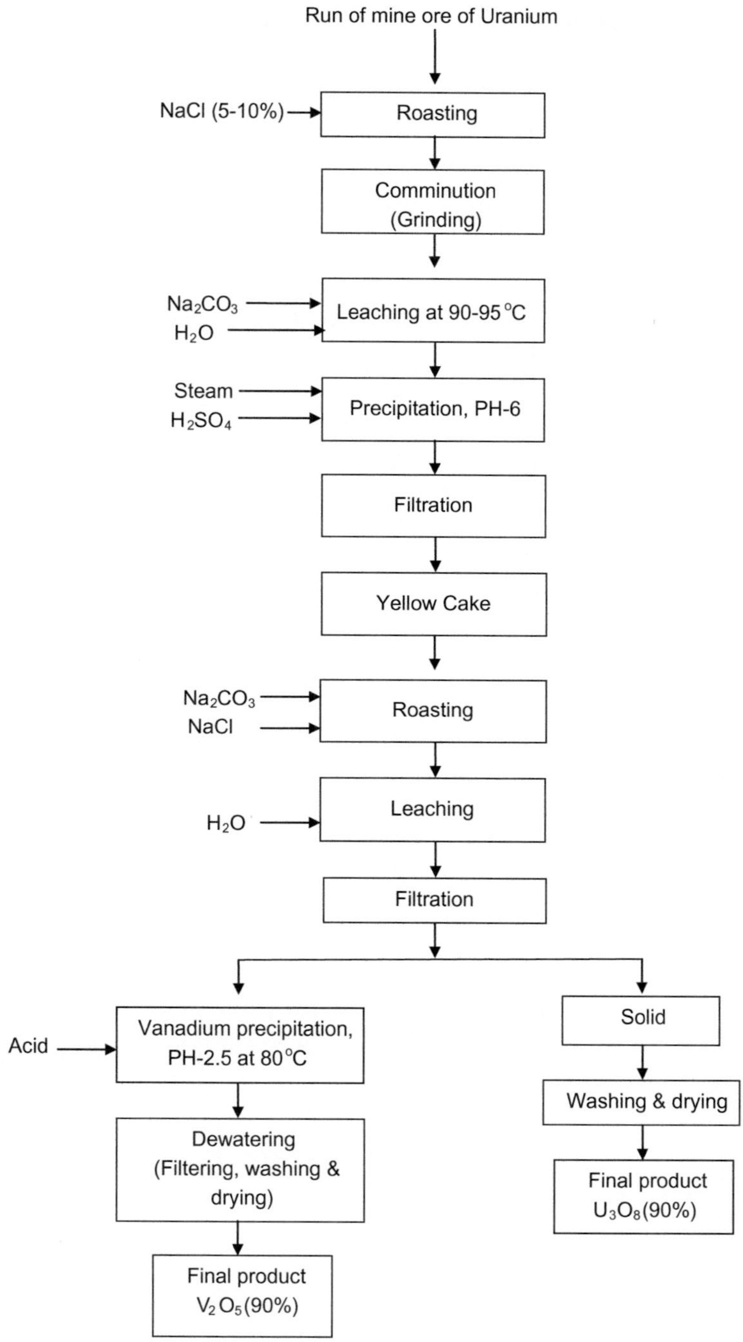

Fig. 19.1 Flow diagram for the recovery of vanadium oxides from uranium ore

(b) *Carbothermic reduction:* The carbothermic reduction in a vacuum has been made possible for all metals of the 5th Group of the Periodic Table (i.e. V, Nb, Ta). The possibilities and conditions for a reduction may be expected on the basis of the Pourbaix-Ellingham diagrams (Fig. 19.2). It should be noted that a number of phase boundaries are pressure dependent, so that for this isobaric presentation, displacements of certain phase fields will result from falling total pressure. Triple points change their temperature with a change in pressure (and vice versa) and correspond to the lines which characterize the four phases of the Pourbaix-Ellingham diagrams.

If it is assumed that no further phases exist, then the carbothermic reduction of vanadium pentaoxide should be interpreted on the basis of the above diagrams as follows:

$$V_2O_5(s) + CO(g) = 2VO_2(s) + CO_2(g) \tag{19.2}$$

$$2VO_2(s) + CO(g) = V_2O_3(s) + CO_2(g) \tag{19.3}$$

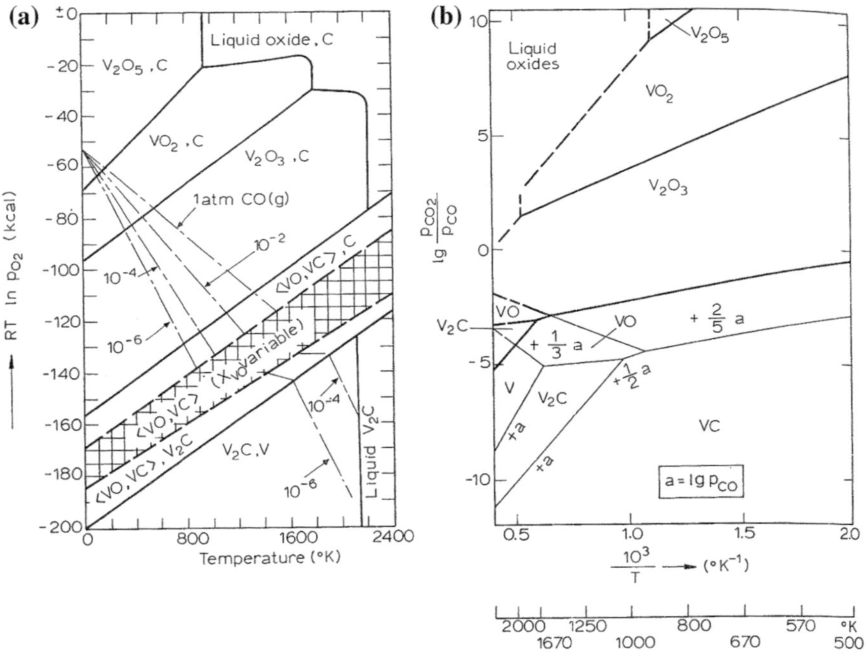

Fig. 19.2 **a** Pourbaix-Ellingham diagram for the system V–O–C. **b** Phase diagram for the system V–O–C for p_{CO} = 1 atm (*heavy line*) and p_{CO} = 10^{-6} atm (*thin line*). *Source* Vacuum Metallurgy, Eds O. Winkler and R. Bakish, Elsevier Pub Co, p. 162

$$V_2O_3(s) + 5C(s) = 2VC(s) + 3CO(g) \tag{19.4}$$

$$2V_2O_3(s) + VC(s) = 5VO(s) + CO(g) \tag{19.5}$$

$$VO(s) + 3VC(s) = 2V_2C(s) + CO(g) \tag{19.6}$$

$$VO(s) + V_2C(s) = 3V(s) + CO(g) \tag{19.7}$$

From these reactions it may be ascertained that neither VO_2 nor V_2O_3 or VC can exist in equilibrium with vanadium. These compounds must be reduced to VO or V_2C before vanadium can form. These complex reactions indicate that partial steps may govern the reaction rate. From Fig. 19.2b, it is evident that for p_{CO} = 1 atm no reduction to the metal can be expected even at 2500 K, but that the range of existence of vanadium metal is extended by a lowering of the pressure, and that in this way technologically feasible reduction conditions may be created. The required minimum reduction temperatures for vanadium, niobium and tantalum oxides are shown in Table 19.1.

Pre-reduced V_2O_3 with lampblack is heated at 900–1000 °C for 2 h. After the conventional intermediate hydrogenation, final reduction takes place at an increasing temperature up to 1700 °C and decreasing pressures down to 10^{-4} torr in five process stages which cover an overall period of 34 h. According to Fig. 19.2b, only VO is obtained in a reduction to V_2O_3 by lampblack at 1450 °C and CO gas pressure above 4 torr, whereas reduction to the metal oxide takes place at 1575 °C and 10^{-2} torr, although again in several process stages.

19.3.3 Ferro-Vanadium

(a) Vanadium content in ore is low and for this reason the manufacture of ferro-vanadium is a combination of a number of metallurgical and chemical processes. Vanadium-bearing iron-titan-magnetite ore (34–38% Fe_2O_3, 28–31% FeO, 13–14% TiO_2, 0.63–0.68% V_2O_3, 4–10% SiO_2, 4–8% Al_2O_3 etc.) is used for smelting pig iron with 0.45–0.55% V. The next stage is the transfer of vanadium from pig iron to the slag in a converter. Nearly 90% V contained in the pig iron is thus reverted to the slag. Vanadium in the slag is in the form of

Table 19.1 Reduction temperatures of the metal oxides of the 5th Group of the Periodic table

Pressure of CO (Atm)	Temperature, K		
	Vanadium oxide	Niobium oxide	Tantalum oxide
1	2700	2780	2860
10^{-3}	2020	2095	2130
10^{-6}	1620	1695	1705

$FeO.V_2O_3$. Evaluated in V_2O_5, the content of V_2O_3 and V_2O_5 in converter slag are average 12–14%.

To recover vanadium from the slag, it is necessary to turn it into compounds soluble in water or in diluted solutions of acids. To this end the converter slag is first crushed and ground in a ball mill; the metallic particles are then eliminated by magnetic separation. The transfer of vanadium to soluble compounds is effected by the oxidization roasting of the slag at 770–820 °C with alkaline additions. Sodium and potassium chlorides are used as alkaline additions, the amount 10–12% of the weight of the slag. The mixture is roasted in a rotating kiln. The bulk of V_2O_3 is oxidized to V_2O_5 and combines with alkaline additions, forming water-soluble compounds: sodium vanadates ($Na_2O.V_2O_5$).

$$V_2O_3 + O_2 = V_2O_5 \tag{19.8}$$

$$2NaCl + H_2O = Na_2O + 2HCl \tag{19.9}$$

$$V_2O_5 + Na_2O = Na_2O.V_2O_5 \tag{19.10}$$

Roasted slag is leached in special installations and insoluble particles are filtered. The filtrate, containing sodium and potassium vanadates, is acidified with sulphuric acid and heated. This results in the precipitation of vanadium pentoxide, which contains 80–90% V_2O_5.

$$Na_2O.V_2O_5 + H_2SO_4 = V_2O_5 + Na_2SO_4 + H_2O \tag{19.11}$$

The precipitate of vanadium pentoxide is filtered, dried and melted at 700–750 ° C. Liquid vanadium pentoxide is poured into a massive iron disc where it solidifies in the form of thin platelets. The recovery of vanadium from the slag is 80–85%. Vanadium pentoxide contains 85% V_2O_5, 2% SiO_2, 6% Fe_2O_3, 1% Cr_2O_3, 1% Al_2O_3, 1.5% MnO, 1% CaO, 0.7% MgO. 0.07% P; the rest is Na_2O and K_2O.

(b) Ferro-vanadium is also produced either by alumino-thermic reduction or by electro-thermic reduction of the V_2O_5 with carbon or silicon. Unless due precautions are taken, loss of vanadium occurs in the slag due to incomplete reduction.

The production of ferro-vanadium by the reduction of its oxides by carbon has not found wide application because vanadium is a carbide-forming element and the resulting alloy, containing 4–6% C, cannot be used for manufacture of low carbon steels. Consequently, the bulk of ferro-vanadium is produced by the reduction of vanadium pentoxide by aluminium or silicon.

$$3V_2O_5 + 10Al = 6V + 5Al_2O_3 \tag{19.1}$$

$$2V_2O_5 + 5Si = 4V + 5SiO_2 \tag{19.12}$$

The reduction with silicon can proceed completely, the silica that is formed is combined with calcium oxide to form 2CaO. SiO_2:

$$2CaO + SiO_2 = 2CaO\ .\ SiO_2 \tag{19.13}$$

Without using calcium oxide, vanadium pentoxide can be reduced partially to vanadium trioxide:

$$V_2O_5 + Si = V_2O_3 + SiO_2 \tag{19.14}$$

Since vanadium trioxide is a basic oxide, it combines with silica to form a silicate and further reduction with silicon becomes impossible.

The raw materials, for the production of ferro-vanadium by the silico-thermic process, are vanadium pentoxide, 75% ferro-silicon, soft iron turning and lime. Smelting is done in two stages: reduction and refining.

In the first period the iron turning is charged into a well-heated bath. As the iron melts down, a mixture of vanadium pentoxide with lime and ferro-silicon (excess of the amount required) is added. As soon as the mixture has melted down, the heat is killed to complete the process. The slag is drained. The silicon content in the alloy at the end of the reduction reaches 15%. After the slagging off, the alloy is refined from silicon by charging of vanadium pentoxide and lime. The vanadium pentoxide is reduced by silicon content in the alloy; as a result, silicon content in the alloy decreases. A small amount of aluminium is added in the reduction period to promote full recovery of the vanadium from the slag.

19.4 Properties

Vanadium is a brittle, very hard metal, light-silvery in colour. Atomic weight of vanadium is 50.95 and it has specific gravity of 6.11, melting point 1900 °C, and boiling point around 3000 °C. Its modulus of elasticity is 12.65–13.36 kg/mm^2. The combination of density and modulus of elasticity indicates that vanadium could be used for structural materials.

It can form a series of solid solutions with iron, an alloy of iron with 31% V having the lowest melting point (1440 °C). With carbon, vanadium can form vanadium carbide (VC) and other carbides. Vanadium carbide has a high hardness and melting point (2800 °C). When heated in atmosphere of nitrogen, vanadium forms vanadium nitride (VN), whose melting point is 2050 °C. Vanadium and silicon form VSi_2 and V_3Si silicides.

Vanadium is stable in dry air at room temperature and oxidizes rapidly at high temperatures. With oxygen, vanadium gives basic oxides VO and V_2O_3, amphoteric oxide VO_2 and acid oxide V_2O_5, their melting points being 1970 °C, 1967 °C,

1545 °C and 675 °C respectively. Vanadium trioxide (V_2O_3) is the most common among vanadium minerals. Vanadium pentoxide (V_2O_5) is used most often as a starting material for smelting vanadium alloys. Vanadium pentoxide reacts with basic oxides to form vanadate salts, such as calcium vanadate ($CaO.V_2O_5$), sodium vanadate ($Na_2O.V_2O_5$) etc.

Since vanadium has a strong tendency to react with oxygen and nitrogen, it is necessary to protect the metal from the atmosphere during hot working. This can be accomplished by heating in an inert atmosphere (purified argon or helium), or the ingot can be covered with a stainless steel jacket. The optimum breakdown temperature range is 1100–1150°C.

19.5 Applications

Vanadium is one of the elements which possess excellent deoxidizing properties. It is also used in the production of special steels. Because of its ability to increase the strength and improve the plastic properties of steel, as well as to raise the later's resistance to attrition and impacts, vanadium is widely used in the production of various structural, tool and spring steels.

As an alloying element in steel, vanadium is used in the form of ferro-alloy (containing 30–40% V). Vanadium is used primarily as an alloying element in special alloy steel. The amount added to steels is generally 0.1–0.25% and in high-speed steels 1–5%. As the carbon-vanadium steels retain their tensile strength well at moderate temperatures, they are used for high pressure boilers and for locomotive parts, while in cast form they find employment as wheel canters, gears and in turbines.

Compounds of vanadium are widely used in various fields like chemical, textile, paint industries, agriculture, medicine, etc. The bulk of vanadium is consumed by steelmaking in the form of ferro-vanadium. Presence of phosphorus, as impurity, in ferro-vanadium can cause difficulties in the manufacture of vanadium-rich steels.

Vanadium pentoxide (V_2O_5) is a powder, orange-yellow or brick-red in colour, with specific gravity of 3.32, melting point of 675 °C and boiling point of 2245 °C. It serves as a raw material for the production of vanadium and ferro-vanadium and is also widely used as a catalyst in the manufacture of sulphuric acid and for hydrogenation in the field of petroleum technology.

Chapter 20
Niobium and Tantalum

20.1 Introduction

Niobium (Nb) belongs to Group V of the Periodic Table (Fig. 1.1). Niobium is a comparatively rare element, its content in the Earth's crust averaging 0.001%. As a rule, it is accompanied by tantalum (Ta) in ores. Columbite (mineral of niobium) was named after Columbia in America, the place of origin of the mineral from which the metal was isolated in 1802. Forty years later, what was thought to be a new element was discovered and named niobium, but further research revealed that niobium and columbium were one and the same element. In Europe the element is still called niobium, whilst in America it is known as columbium.

Tantalum was discovered in 1802 by Ekeberg of Sweden, he named it after Tantalus in Greek mythology because of the difficulty of dissolving the oxide.

20.2 Sources

Niobium has not been found in nature in its free state; it always occurs as an oxide, together with tantalum, and is associated with other minerals having calcium, iron, manganese, and rare earths metals as bases. Niobium occurs associated with tantalum in the mineral columbite [(Fe, Mn)O.(Nb, Ta)$_3$O$_5$], which has been considered the most important source of niobium. Columbite deposit is found in Africa. Tantalum occurs associated with niobium in tantalite (Ta_2O_5.FeMnO). Tantalite deposites are found in USA, Brazil, Australia, and South West Africa. Columbite and tantalite are in reality end members of an isomorphous series of tantalates and columbates, one mineral merging with the other. The minerals are black in colour, with a specific gravity ranging from 5.3 for columbite to 8.0 for tantalite, the density of a concentrate thus is some indication of the ratio of the respective minerals.

S.K. Dutta and D.R. Lodhari, *Extraction of Nuclear and Non-ferrous Metals*,
Topics in Mining, Metallurgy and Materials Engineering,
https://doi.org/10.1007/978-981-10-5172-2_20

Pyrochlore, a fluo-niobate of calcium and sodium ($NaCaO.Nb_2O_5F$) is a potential source of the niobium metal. Vast low-grade (0.2–0.6% Nb_2O_5) deposits occur in Tanzania, Uganda, Canada, Africa and Norway.

20.3 Extraction

20.3.1 Separation of Niobium and Tantalum from Ores

The flow diagram for extraction of niobium and tantalum from a columbite-tantalite ore is shown in Fig. 20.1. Recovery and separation of tantalum from niobium is based on the difference of solubilities of the double potassium fluorides. Sodium tantalite and niobate, first formed by fusion with caustic soda and soda ash, are then washed with water to remove soluble TiO_2, SiO_2 and SnO_2 formed by the fusion. Leaching with 50% sulphuric acid obtains the corresponding tantalic and niobic acids. After settling, soluble Fe and Mn are removed by decantation, and the residue is washed with hydrofluoric acid. Addition of potassium fluoride leads to the formation of the double fluorides (K_2TaF_7 and K_2NbF_7) which crystallize out on standing. Recrystallization from dilute hydrofluoric acid yields substantially pure potassium tantalum fluoride. Potassium niobium fluoride remains in the solution from the recrystallization; then from the solution, Nb_2O_5 is obtained by precipitation with ammonia. Dissolution and precipitation is done to pure potassium tantalum fluoride residue to obtain Ta_2O_5.

20.3.2 Metallic Niobium

Production of niobium metal is done by carbothermic reduction of niobium pentoxide. The carbothermic reduction has been carried out on an industrial scale for a number of years. The reduction reactions are as follows:

$$Nb_2O_5(s) + C(s) = 2NbO_2(s) + CO(g) \tag{20.1}$$

$$Nb_2O_5(s) + 7C(s) = 2NbC(s) + 5CO(g) \tag{20.2}$$

$$3Nb_2O_5(s) + NbC(s) = 7NbO_2(s) + CO(g) \tag{20.3}$$

$$2NbO_2(s) + 10NbC(s) = 6Nb_2C(s) + 4CO(g) \tag{20.4}$$

$$3NbO_2(s) + Nb_2C(s) = 5NbO(s) + CO(g) \tag{20.5}$$

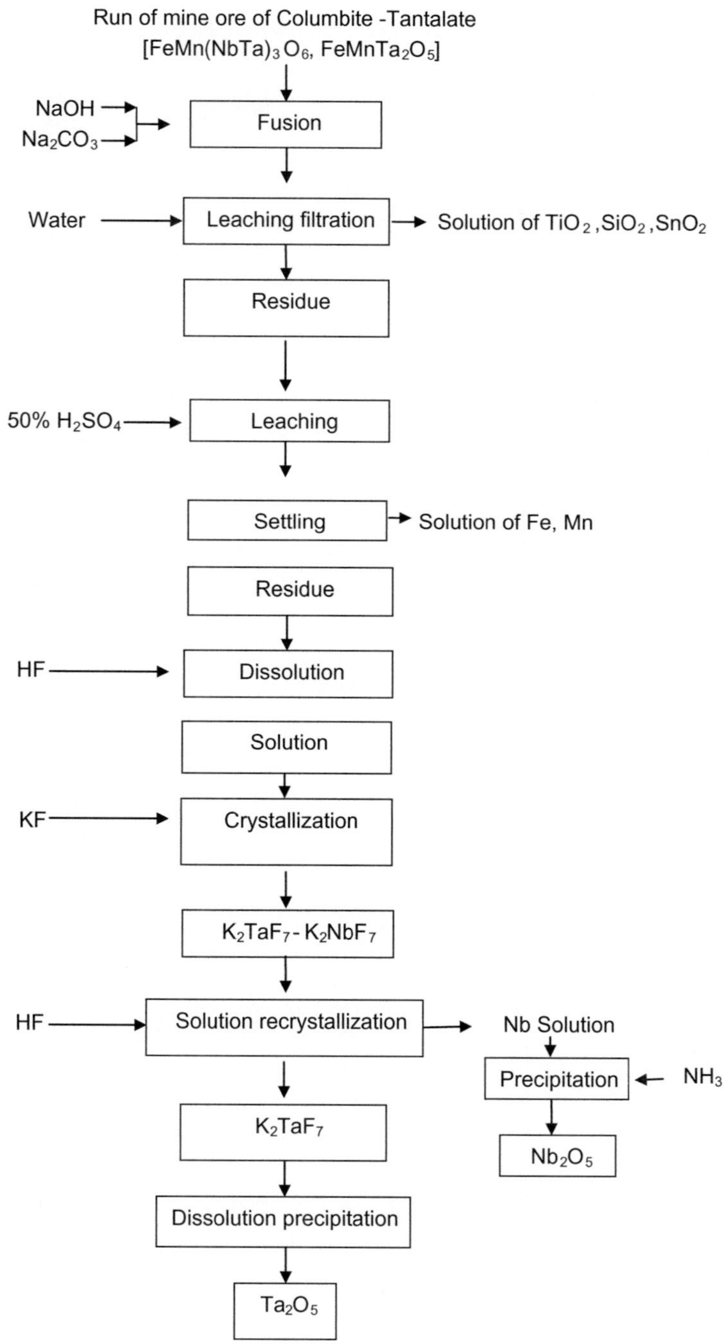

Fig. 20.1 Flow diagram for recovery of niobium and tantalum

$$NbO(s) + Nb_2C(s) = 3Nb(s) + CO(g) \tag{20.6}$$

From these equations it may be inferred that neither NbO_2 nor NbC but only NbO and Nb_2C are stable in the presence of Nb. No reduction by carbon is possible at atmospheric pressure up to 2500 K (as shown in Table 20.1), but by applying ultimate pressures of 10^{-3} torr technologically suitable working conditions are reached. The technological process of the production of Nb is usually carried out as a three-stage process:

(1) After production of NbC, compacts of Nb_2O_5 and NbC are prepared, in which an O:C ratio of about 1.15 should be maintained.
(2) In the following vacuum reduction several working stages are distinguished as is shown in Table 20.1. While the loss in weight in this second stage is attributable to a marked degree to liberation of CO, the loss in weight in the third stage should be attributed mainly to the evaporation of the sub-oxides. The reduction is usually carried out in induction heated vacuum furnaces with high pumping speeds.
(3) After an intermediate hydrogenation, crushing and adjustment of composition, a post refining by sintering or melting in a high vacuum is carried out.

$$Nb_2O_5(s) + C(s) = 2NbO_2(s) + CO(g) \tag{20.1}$$

$$2NbO_2(s) + 2C(s) = 2NbO(s) + 2CO(g) \tag{20.7}$$

$$2NbO(s) + 2C(s) = 2Nb(s) + 2CO(g) \tag{20.8}$$

From X-ray analysis it was found that NbO could be identified in the reaction (20.7) only after a temporary appearance and disappearance of NbC. In the reaction (20.8), NbC was formed as primary product, and was then converted to Nb_2C. Only after the disappearance of the NbC was it possible to identify Nb besides Nb_2C and NbO. For final pressures of 10^{-3} torr, agreement with the thermodynamic considerations is excellent. The reduction has begun at 1050 °C to a marked extent if lampblack instead of NbC is used, and that it is almost complete at 1730 °C.

Table 20.1 Reduction of Nb_2O_5 by NbC to niobium in a vacuum

Working range	Temp (°C)	Final pressure (Torr)	Time (Min)	Analysis (wt%)			Weight loss (%)
				O	C	N	
Starting mixture	–	–	–	10	7	–	
1	1500	10^{-2}	80	6	2	0.39	
2	1800	10^{-3}	100	2.5	1.1	0.41	12
3	2100	10^{-4}	140	<0.1	<0.2	0.02	16

The niobium pentoxide (Nb_2O_5) is reduced with carbon according to the overall reaction:

$$Nb_2O_5(s) + 5C(s) = 2Nb(s) + 5CO(g) \quad (20.9)$$

As per reaction (20.9) above (2 × 92.9) kg of Nb is produced along with (5 × 22.4) m^3 of CO gas at one atmosphere (760 torr) pressure; i.e. 0.6 m^3 of CO gas, per kg Nb produced, must be pumped off if a pressure of 760 torr prevails in the reaction vessel; but at a pressure of 10^{-3} torr, 456,000 m^3 of CO gas must be handled. This means that such processes must be divided into several individual steps in order for the large quantities of gas to be pumped off at the highest possible pressure. The major portion of the gas formed is taken out at atmospheric pressure, so that under high vacuum conditions only a relatively small quantity of gas has to be removed. The niobium powder is finally produced.

The niobium powder must be consolidated into ingot form by sintering. Briquettes are made at the pressure of 23.6–31.5 kg/mm^2. Briquettes are sintered at a temperature of 1700–2300 °C and under vacuum. Major purification is done during sintering. Hydrogen is rapidly eliminated by out-gassing at 500–1000 °C. Carbon and oxygen are removed at above 1400 °C by combining to form carbon monoxide, which diffuses to the metal surface and goes out. Finally, nitrogen is removed by simple out-gassing at above 1900 °C.

Further purification of the niobium metal is done by vacuum drip-melting using either high frequency induction or electron beam refining methods. The melting of niobium is rendered difficult by their affinity for carbon, oxygen and nitrogen. Several methods have been developed which avoid or minimize such contamination. One of the methods is the drip-melting process, in which a vertical rod of the material is gradually lowered down into the field of an induction coil, so that the lower portion of the rod is continuously melted. The molten drops then fall into a water-cooled copper crucible.

Heating by electron bombardment in a vacuum of 10^{-3} mm of Hg or less has been used for drip-melting of niobium. The impure niobium ingot is fed into the vacuum chamber from the top; high voltage electrons are supplied by a horizontal DC power source, and are focused on the lower portion of the ingot. As the ingot melts, the drops of metal fall into a molten pool from where continuous withdrawing forms the purified niobium ingot. Unwanted impurities thus are largely volatilized during melting.

20.3.3 Metallic Tantalum

Metallic tantalum can be produced by the carbothermic process in a vacuum. According to published reports, only oxide phase Ta_2O_5 exists, whereas other reports mention the existence of TaO and TaO_2. The two systems are compared in Fig. 20.2. Accordingly, the following equilibrium reactions occur:

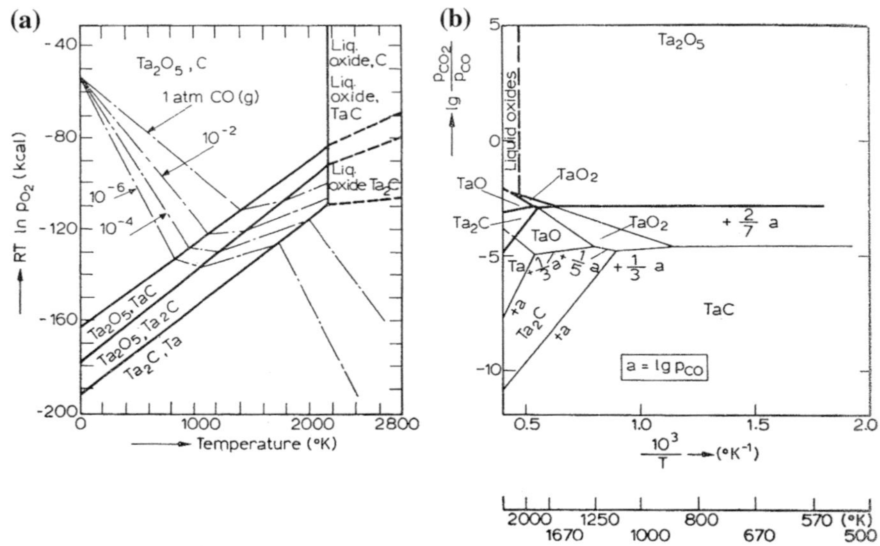

Fig. 20.2 **a** Pourbaix-Ellingham diagram for the system Ta–O–C with Ta_2O_5 as the sole oxide. **b** Thermodynamic equilibrium diagram for the system Ta–O–C for $p_{CO} = 1$ atm (*heavy line*) and $p_{CO} = 10^{-6}$ atm (*thin line*) with the oxide phases Ta_2O_5, TaO_2 and TaO. *Source* Vacuum Metallurgy, Eds O. Winkler and R. Bakish, Elsevier Pub Co, p. 174

$$Ta_2O_5(s) + 7C(s) = 2TaC(s) + 5CO(g) \tag{20.10}$$

$$Ta_2O_5(s) + 12TaC(s) = 7Ta_2C(s) + 5CO(g) \tag{20.11}$$

$$Ta_2O_5(s) + 5Ta_2C(s) = 12Ta + 5CO(g) \tag{20.12}$$

In the reduction of Ta_2O_5 by C or TaC at 1200 °C in a vacuum, only Ta_2O_5 and carbide phases (Ta_2C, TaC) were found in accordance with the phase diagram. Although TaO and TaO_2 should be found, their existence does not seem to be certain at low temperatures. The rate of reaction with carbon is determined by the history of the oxide, the gas pressure and the contact between particles; the rate of reaction with TaC is determined by the shape of the particles, the density of the charge and the pressure of the gas phase.

From the experiments it emerged that it is advantageous to maintain an O:C ratio of 1.03, but this ratio had no marked effect on reduction of Ta_2O_5 above 1700 °C; neither the reduction by CO with temporary formation of CO_2 (as shown in Fig. 20.2b) nor a solid state reaction would be expected to play a role. Diffusion of C and O in the metal as well as the desorption of TaO or TaO_2 respectively, and subsequent reactions with Ta_2C, are more likely to determine the rate of reduction. The pure metal may be produced without any difficulty at temperatures between 2000 and 2700 °C. Finally the tantalum powder is produced.

Because of the high melting point of tantalum and its great chemical reactivity when heated, special techniques must be used for purification and to convert it to massive form. In general they are similar to niobium. The methods used industrially are sintering, vacuum or inert atmosphere arc melting, and electron beam melting.

20.3.4 Ferro-Niobium

The production of a Nb-containing ferro-chromium addition for steelmaking is possible at 1500 °C. In this process, high carbon, crushed (0.07–0.037 mm in diameter) ferro-chromium is mixed with niobium pentoxide Nb_2O_5 to an O:C ratio of 1–1.2, compacted and treated below 1500 °C in a vacuum (below 3 torr). Oxygen and carbon are almost completely removed by this process.

When niobium pentoxide (Nb_2O_5) is reduced by silicon, incomplete reduction occurs to lower oxides along with its reduction to metal. Further reduction of lower oxides by silicon is difficult and leads to the formation of silicon-bearing alloys, as well as high niobium losses in slag.

Production of ferro-niobium is done by alumino-thermic reduction of niobium pentoxide in an electric arc furnace. The charge consists of concentrate of niobium [40–50% $(Nb + Ta)_2O_5$, 12% SiO_2, 8% TiO_2 etc.], aluminium powder, iron ore and sodium saltpeter ($NaNO_3$). The iron ore should contain at least 60% iron and not more than 4% silica and 0.15% phosphorus. Sodium saltpeter contain 98% $NaNO_3$. The charge is preliminarily mixed in a drum mixer and supplied into the a bin above the furnace. Charging is done through charging hole. Smelting is carried out at a voltage of 140 v. Upon melt-down, a holding time of 1–2 min is allowed, and then a small amount of aluminium powder is added to recover niobium from the slag. The highest niobium recovery is attained with the use of aluminium as a reducing agent. The chief efficiency factor in ferro-niobium smelting is the recovery of niobium, for that cost comes to about 90% of the total production cost. Yield of niobium is about 96–98%. The ferro-alloy contains about 50–70% Nb and Ta. The main efficiency factor in ferro-niobium smelting is the recovery of niobium, for its cost comes to approximately 92% of that of alloy production.

The reactions of ferro-niobium smelting by alumino-thermic reduction of niobium pentoxide are as follows:

$$3Nb_2O_5 + 10Al = 6Nb + 5Al_2O_3 \tag{20.13}$$

$$3Ta_2O_5 + 10Al = 6Ta + 5Al_2O_3 \tag{20.14}$$

$$3SiO_2 + 4Al = 3Si + 2Al_2O_3 \tag{20.15}$$

$$Fe_2O_3 + 2Al = 2Fe + Al_2O_3 \tag{20.16}$$

$$3FeO + 2Al = 3Fe + Al_2O_3 \quad (20.17)$$

$$3TiO_2 + 4Al = 3Ti + 2Al_2O_3 \quad (20.18)$$

20.4 Properties

20.4.1 Niobium

Elemental niobium is a plastic metal, grey in colour, with atomic weight of 92.91, specific gravity of 8.57; melting point of 2415 °C and boiling point of 3300 °C. With iron, niobium gives a compound Fe_3Nb_2 which may dissolve a great amount of iron but not in niobium. With carbon, niobium forms carbides of varying composition from NbC to Nb_4C.

With oxygen, niobium forms three oxides: niobium pentoxide (Nb_2O_5), niobium dioxide (NbO_2) and niobium monoxide (NbO). Niobium pentoxide is an acid oxide and a fine crystalline powder, white in colour, with specific gravity of 4.55, melting point of 1460 °C. Niobium dioxide is a powder, blue-black in colour; it is unstable and decomposes to Nb_2O_5 when heated. The most stable oxide is the niobium monoxide. It is a black powder with specific gravity of 6.27. The character of niobium oxides varies from acid to basic with a decrease in the amount of oxygen linked to niobium.

20.4.2 Tantalum

Tantalum, atomic weight 180.95, is located in Group VB of the Periodic Table (Fig. 1.1) below its sister element niobium, and adjoined by hafnium on the left and tungsten on the right. It is a strong ductile metal characterized by: (a) its high specific gravity, 16.6; (b) its high melting point, 2996 °C, the third highest among the metals, exceeded only by rhenium (Re), 3170 °C, and tungsten, 3410 °C; (c) its high boiling point, 6000 °C, the highest among the metals; (d) the tenacious thin oxide layer on its surface which gives it superior rectifying and capacitance properties; and (e) its extreme inertness to attack by all acids, except hydrofluoric and fuming sulphuric, at ordinary temperatures.

Tantalum is a strong, ductile, hard metal somewhat comparable to mild steel. Although working improves its hardness and tensile strength, tantalum becomes work hardened much more slowly than most metals; thus high reductions may be possible between anneals. The presence of impurities likes C, O, N, and H greatly affect the mechanical properties of the metal.

Tantalum is one of the most inert of all metals to reaction with chemicals at temperatures below 150 °C. The only reagents which attack it rapidly are fluorine, hydrofluoric acid. As the temperature of exposure becomes higher, tantalum begins to react slowly with a large variety of reagents, the rate of attack increasing with the temperature rise.

Tantalum and niobium, two corrosion resistant metals, form alloy in all proportions. The presence of even small amounts of niobium, like <5%, in tantalum greatly reduces the corrosion resistance of the later. The corrosion rates increase roughly in proportion to the niobium content of the alloys, and even though the 95Ta–5Nb alloy shows excellent resistance in all exposures, the attack is three times that obtained on pure tantalum. Tantalum and tungsten alloy containing more than 18% tungsten show zero corrosion rate in 20% HF, an advantage over tantalum.

20.5 Applications

20.5.1 Niobium

The addition of niobium to stainless and heat-resisting steels increases their plasticity and corrosion resistance. The introduction of niobium in structural steels improves their welding properties.

Niobium metal offers an excellent resistance to corrosion at elevated temperatures. Niobium alloys are ideal for high temperature applications, because they are neither affected by creep nor corroded by hot gases. Ferro-niobium is used for special steelmaking. Niobium increases the strength of extra low carbon steels at elevated temperatures. Percentages of niobium present in alloys for high temperature work vary from 0.4 to 1.0%.

Niobium is used as a hardening addition to the cobalt base, nickel base and chromium base alloys at high temperature. Because of its excellent performance at elevated temperatures, its low capture cross-section for thermal neutrons and its resistance to attack by liquid metals, niobium can be used as fuel containment for high temperature nuclear reactors.

Tantalum metal has certain noteworthy properties, such as ductility, a high melting point, a high strength at high temperatures, and resistance to corrosion by acids, bromine, and chlorine at normal temperatures. At high temperatures, tantalum is corroded by alkalies and hydrofluoric acid.

Niobium and tantalum have high melting points, i.e. 2415 °C and 2996 °C, respectively. These two metals along with titanium, zirconium, tungsten and molybdenum are known as refractory metals, with melting points in the range of 1800–3000 °C. Some of them are used in the manufacture of jet engines and guided missiles.

20.5.2 Tantalum

Applications of tantalum are based on its high melting point, good strength and ductility, low vapour pressure, inertness to chemical attack at below 150 °C, getting properties at elevated temperatures.

The largest use of tantalum is in electrolytic capacitors. A tantalum electrolytic capacitor has four components: (a) the tantalum anode (foil, wire, or powder sintered); (b) the tantalum oxide dielectric surface (anodic oxidation of the surface of the tantalum anode); (c) the electrolyte which acts as the cathode (liquid [solutions of sulphuric or phosphoric acid], or solid [semiconductors such as higher oxides of manganese, lead or nickel]); and (d) the container or coating in contact with the electrolyte and chemically inert to it.

Fabrication of corrosion resistant equipment is perhaps the second largest application of tantalum. In addition to it being strong and essentially inert to attack by non-alkaline highly corrosive media at ordinary temperatures (with the exception of F_2, HF and free SO_2), tantalum is characterized by extremely high heat transfer coefficients. Tantalum is the most like glass as a corrosion resistant material and it is often used with glass, glass-lined steel, and other non-metallic materials for chemical equipment.

Tantalum is used to construct equipment for the preparation and handling of molten metals, such as the rare earth metals, and alloys containing fissionable materials, at elevated temperatures. For these uses tantalum must be protected from exposure to the atmosphere.

The complete immunity of tantalum to human body fluids and its non-irritating tolerance by body tissues have resulted in its wide use for surgical repairs. Tantalum is applied as plate and sheet in bone repair (specially in cranial sites), as wire for sutures, as foil and wire for nerve repair, and as plate, sheet, and woven gauze for abdominal muscle repair. Tantalum powder is used for covering the side of an excised brain tumor, before the skull is closed.

While tantalum's thermal neutron absorption cross-section is too high to permit it to be used in or near the core of most types of atomic reactors, it can be used in fast neutron reactors. Tantalum is used for handling high temperature liquid metals, such as sodium or NaK, in heat transfer equipment, and for liquid metal systems such as the bismuth-uranium alloy.

Tantalum is used for components in high temperature furnaces operated under vacuum or inert atmosphere. Tantalum oxide is used as an ingredient in some kinds of high refractive index optical glasses. Tantalum carbide is used in cutting tools where it imparts shock resistance and a very low coefficient of friction.

Part IV
Production of Ultra-High Purity Metals

Chapter 21
Methods of Refining

21.1 Introduction

The advance of atomic energy, electronics etc. manufacture has led to a demand for metals and alloys of a very high order of purity which has necessitated the development of new operational techniques. Most of the nuclear reactor metals are very sensitive to the presence of impurity. Hence, these metals require producing ultra-high purity for use as nuclear reactor metals. Ultra-high purity of metals can be achieved by the following methods:

(1) Zone refining.
(2) Vacuum induction melting.
(3) Vacuum arc melting.
(4) Inert atmosphere arc melting.
(5) Electron beam melting.

21.2 Zone Refining

The advancement of the semiconductor industry with its requirement for ultra-high purity materials led to the development of zone refining. Zone refining (melting) denotes a method of controlling the distribution of soluble impurities or solutes in the crystalline materials which are partially melted in a particular zone. This technique consists of melting a small zone of the bar by induction heating, and then moving the melted zone along the length of the bar by a corresponding movement of the source of heat. The result is a sweeping of impurities towards one end of the bar due to the greater solubility in the liquid phase of the impurities. In zone refining, a small molten zone moves through a long charge of alloy or impure material. The differences in impurity concentration between liquid and crystallizing

S.K. Dutta and D.R. Lodhari, *Extraction of Nuclear and Non-ferrous Metals*,
Topics in Mining, Metallurgy and Materials Engineering,
https://doi.org/10.1007/978-981-10-5172-2_21

solid lead to segregation of impurities into the moving liquid zone and a corresponding purification of the solidifying material. A molten zone, during travelling, has two liquid–solid interfaces; i.e. a melting interface and other freezing interface. The zone is able to redistribute impurities mainly because of the freezing interface. The main difference between zone melting and fractional crystallization is that one part of the charge is molten at a time in zone melting, where all the charge is molten during fractional crystallization.

Zone refining is generally used to produce ultra-pure metals. The principle of zone refining may be understood with the phase diagram (Fig. 21.1). When a liquid is cooled slowly, then solidification begins at a particular temperature. The initial solid is purer. As the temperature is reduced, more and more solid separates, the final solid formed having less impurity. The impurity concentration is lower in the solid than in the liquid at equilibrium. The differences in the impurity concentration between liquid and crystallizing solid led to segregation of impurities into the moving liquid zone and corresponding purification of the solidifying material. Impurities of solid move along with solidus line and impurities of liquid moves along with liquidus line.

With the help of a heating device (Fig. 21.2), a molten metal zone is obtained at one end of the metal bar; then the molten zone is slowly moved from one to the

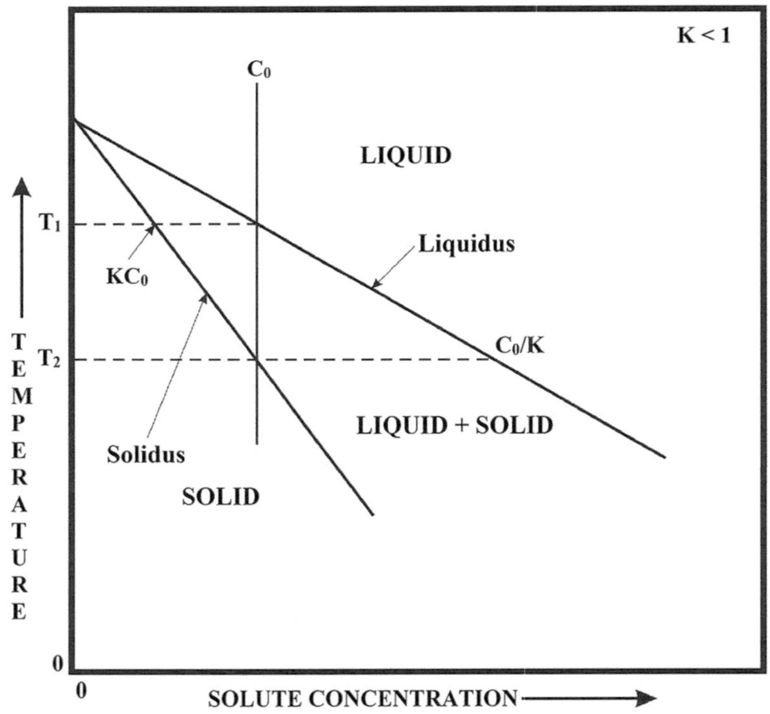

Fig. 21.1 Phase diagram for solidification

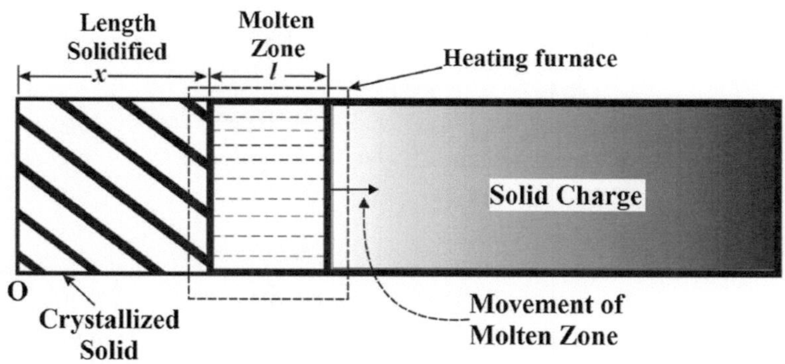

Fig. 21.2 Schematic diagram of zone refining

other end of the bar, by moving the heating device. Thus, every time, impurities are thrown into the liquid zone and hence pure solid is separated from the impure liquid. This process is repeated a few times to get ultra-pure metal. The impurities are finally segregated at one end of the bar which is chopped out.

Refining by solidification is possible because impure metals solidify over a range of temperature with significant difference in the equilibrium composition of liquid and solid (Fig. 21.1). For a liquid metal of initial composition C_o, the first solid to form would be of composition kC_o. The constant k is defined by the ratio of the equilibrium concentration of solute in the solid at the liquid–solid interface of the concentration of the liquid. The distribution coefficient, k, is less than unity, and the liquid would be enriched in solute during solidification.

21.3 Vacuum Induction Melting

The vacuum induction melting process is mostly used for re-melting of virgin metals under vacuum or inert atmosphere, to avoid oxidation and contamination from the atmospheric oxygen and nitrogen. The molten charge is then refined and subsequently cast under vacuum. The product may be in the form of ingot for subsequent processing like forging, rolling etc. A major use of vacuum induction melting is the production of very long electrodes (e.g. 4.6 m) for vacuum arc re-melting. For small units it is not uncommon to hand-charge the crucible under atmospheric conditions, on larger units it is the normal practice to operate with the melting chamber under continuous vacuum and to bulk charge through a vacuum lock. This lock is normally situated directly above the crucible and has a diameter at least as great as that of the crucible. The material to be charged, either virgin metal or alloy, is contained in a drop-bottom basket.

As one limitation in the heat cycle is the refining time, governed solely by metallurgical considerations, and as this may take several hours, an overall heat

Table 21.1 Typical power supplies for vacuum induction melting units

Power supply	Furnace size			
	80 kg	300 kg	1 tonne	5 tonne
Power (kW)	100	300	600	1600
Voltage (v)	250	400–600	400–600	400–600
Frequency (kHz)	2–4	1–2.5	0.5–1	0.15–0.5

time of 4–6 h is normally considered acceptable. This means that the melt-down time must be limited to about 2 h. Therefore, it is important to ensure that furnaces are adequately power supplied. Some typical power supplies for various sizes units are shown in Table 21.1. As the furnace sizes are increased the current frequencies are decreased with increasing power.

After proper melting, it is necessary to be able to sample the liquid bath at different stages of the melt to determine the correct analysis. Small metallic or refractory moulds are introduced at the end of the rod through a small vacuum lock. To achieve maximum refining [main impurities of interest to the metallurgist are the gases (O_2, N_2, H_2), certain non-metallic impurities (S and P), and metallic impurities (Pb and Sn)], prolonged exposures to vacuum of less than 10 mTorr are required. It is normal to specify equipment capable of an ultimate vacuum over liquid bath in the pressure range 1–10 milli torr.

Bottom pour furnace has a water-cooled bottom section or chamber, which contains the mould or mould assembly in which the ingot is cast and ports for connecting vacuum equipment. To pour the molten metal into the mould there is a mechanical device for pulling a stopper rod from the top. Bottom pour furnaces have the advantages of being simple and inexpensive. They are especially suited to the use of graphite crucibles, which minimize the problem of maintaining and controlling the pouring hole closer as compared to ceramic crucibles. This technique is mostly used for uranium and zirconium metals. The compositions of some typical refractories used for pre-fired crucibles are shown in Table 21.2.

21.4 Vacuum Arc Melting

This technique is developed mainly for consolidating and refining reactive and refractory metals (e.g. Ti, Zr, W etc.) which are initially produced in the form of powder, sponge or granules. In this process, heat is generated by the application of a low voltage, high current electric arc to the surface of the molten metal from an electrode which may be either consumable or non-consumable. In the later case, the electrode is to be inert and plays no part in the melting other than to supply heat for the fusion of the raw material. This electrode is usually of a highly refractory metal such as tungsten or of graphite or metal carbide, and in order to minimize vaporization of the electrode, melting is usually carried out under a low pressure of inert gas or in a high vacuum.

Table 21.2 Typical refractory used in vacuum induction melting

Application	Material	Chemical analysis (%)					
		SiO_2	Al_2O_3	Fe_2O_3	CaO	MgO	ZrO_2
Rammed furnace linings	Magnesia	Tr.	Tr.	Tr.	Tr.	99	–
	Commercial Mix (USA)	3.5	20	Tr.	Tr.	75	–
	Commercial Mix (UK)	5.0	26	Tr.	Tr.	68	–
Pre-fired crucibles	Alumina	4.0	95	Tr.	Tr.	Tr.	–
	Magnesia	2.0	Tr.	1.0	Tr.	96	–
Sample moulds	Zircon	34.0	Tr.	Tr.	Tr.	Tr.	65
Pouring tundishes	Firebrick	56.0	40	2.0	Tr.	Tr.	–
	Sillimanite	41.0	57	1.0	Tr.	Tr.	–
	Zircon	32.0	1	1.0	Tr.	Tr.	65

The consumable vacuum arc melting furnace is used extensively for melting thorium and certain uranium alloys (particularly Zr–U). The consumable electrode, made up of the material to be melted, is melted by an arc maintained between the molten pool of metal and the bottom of the electrode. This type of melting eliminates the contamination of metal and provides a high yield of metal. Crucibles are made from copper tube welded to a flange which together with the water jacket is clamped to the furnace body to position the crucible and form the vacuum seal. This flange acts as the anode and carries the full melting current; the electrode acts as the cathode to complete the electrical connection. Earlier crucibles had a solid integral bottom necessitating inversion to remove the ingot, but the modern tendency is towards removable bottoms.

The consumable vacuum arc melting furnace normally operates in the range from about 1 to 10 milli torr and the vacuum system utilized must fulfill two requirements: (a) to evacuate the gases within the furnace; and (b) be capable of handling the gas evolved during the melting of the electrode.

Figure 21.3 shows a consumable vacuum arc melting furnace. If the impure metal is in the form of powder, it is pressed into compacts of suitable size which are then welded together to form a long electrode. Impure sponge metal is cast or forged in the form of an electrode. A direct current (d.c.) power supply is used for arc melting. The impure metal electrode acts as the cathode, and a metal pad kept on the base of the water-cooled copper crucible acts as the anode. After the whole furnace chamber is evacuated (at 10^{-3} mm of Hg pressure), the electric arc is struck between the impure metal electrode (cathode) and metal pad (anode). Due to arcing, re-melting take place at the impure metal electrode; gases present in impure metal and volatile impurities are removed at the vacuum. The impure metal electrode is gradually lowered with the help of sliding vacuum seal as its tip melts due to arcing. Constant spacing is maintained between the electrode tip and the molten pool which is formed below the electrode. Arc control during melting is essential for

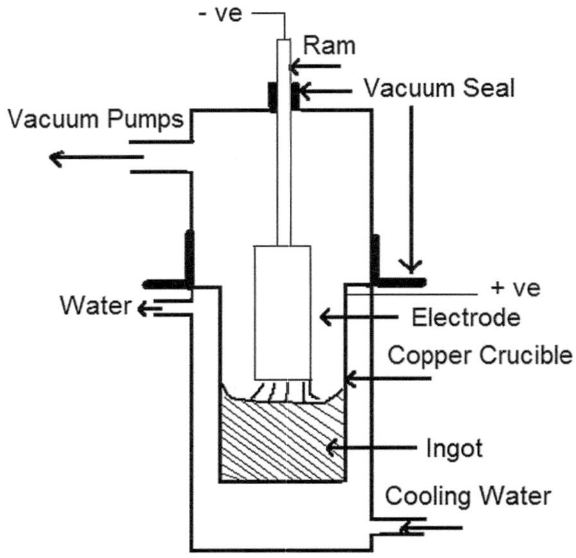

Fig. 21.3 Vacuum arc furnace

homogeneity of the resulting ingot. The melting current may automatically be disconnected if the furnace pressure rises above a pre-set level or in the event of a water failure to the cooling system of the vacuum pumps.

Because the arc is an electrical phenomenon comprising flowing electrons and ionized particles, it may be influenced and deflected mechanically by magnetic fields, intentional or otherwise. The effects of electromagnetic stirring or arc stabilizing may be as follows:

(1) Intentional stirring and agitation of the molten pool.
(2) The avoidance of undesirable rotation of the pool's melting zone.
(3) Alteration of the heat distribution within the melting zone.

A rule of thumb for current input vs mould diameter in a consumable arc melting furnace is that approximately 236 amp/cm of mould diameter is required for melting in the range of 1200–2200 °C. Approximately 2500 amp is required to produce 130 cm diameter. For Zr ingot, 6000 amp melts a 150 cm electrode into a 250 cm crucible at approximately 4.54 kg/min.

Again, since the arc is a constant voltage phenomenon, the melt-off rate of material from the electrode is proportional to current and may conveniently be expressed in terms of weight per kilo ampere-minute. Some typical values are shown in Table 21.3.

During melting, the following metallurgical effects are observed:

(1) There is considerable degassing in the arc zone, mainly from the electrode and molten bath surface, and volatile impurities are removed depending on the pressure and temperature of the chamber.

Table 21.3 Typical values for specific melt-off rates

Serial No.	Material	Kg/kA min
1.	Iron and nickel base alloys	0.565–0.70
2.	Niobium and molybdenum alloys	0.255–0.32
3.	Titanium (sponge)	0.53
4.	Titanium (remelt)	0.55
5.	Copper	0.70

(2) Dissolved gases in the impure metal are removed by diffusion from the molten metal-gas interface: $2\underline{H}/2\underline{N} = H_2/N_2$ (g). The gases are continuously pumped out from the chamber.
(3) Deoxidation of impure metal takes place, if an element (which has strong affinity for oxygen) is present in the metal; e.g. if carbon is present in a tungsten electrode, both carbon and dissolved oxygen are removed: $\underline{C} + \underline{O} = CO$ (g).
(4) The cleanliness of an ingot depends on the size, distribution, and concentration of the non-metallic inclusions. As the metal droplets falls from the electrode to the molten pool, large inclusions are dispersed by the arc which form uniform distribution of small inclusions throughout the melt.
(5) The vacuum remelting process produces an ingot which is free from macro-segregation. When a large volume of a liquid solidifies, the first solid that freezes may have a different composition from the last drop that freezes. This effect is minimized in the vacuum arc melting process.
(6) The vacuum arc melting process does not remove sulphur and phosphorus, which may be present in the impure electrode.
(7) The water-cooled copper mould helps to freeze the pure metal faster.

21.5 Inert Atmosphere Arc Melting

Many metals have great affinity towards oxygen, nitrogen, hydrogen and carbon. Some of them are obtained from the initial reduction stage in the form of finely divided powders of higher surface area, and reactivity requires the use of special techniques for the formation of massive metal forms. Any reaction between the metal and the surrounding gas can be prevented when melting of metal is carried out in an inert atmosphere provided by gases like argon or helium. Tungsten, molybdenum, tantalum, niobium, titanium, zirconium and hafnium are melted in inert atmosphere arc furnace to form massive ingots. A compacted bar of the metal powder or sponge is fed as a consumable electrode into a furnace maintained an inert atmosphere of helium or argon, the molten metal being collected in a

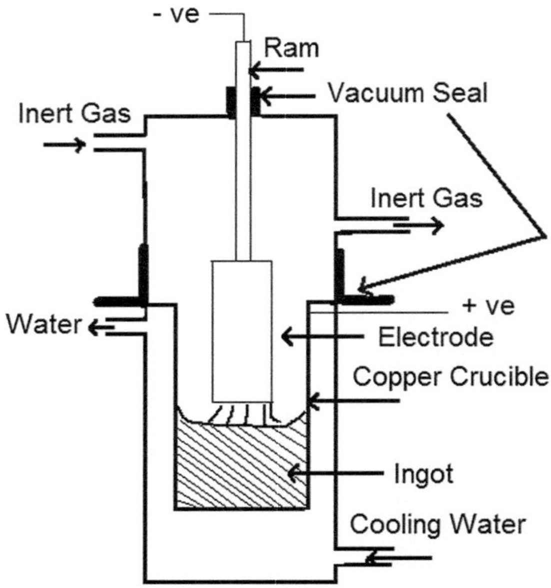

Fig. 21.4 Inert gas arc furnace

water-cooled copper crucible which serves as the second electrical terminal. There is a chance of pick-up of impurities from a ceramic crucible, that is why a copper crucible is used.

Figure 21.4 shows the inert gas arc furnace. A non-consumable tungsten (W) electrode can also be used to strike the arc for melting, in which the metal is fed directly into the molten pool on the lower electrode. An arc is struck between the upper electrode and some pieces of metal (which will be melted) on the base of the water-cooled copper mould, and the upper electrode is advanced as it is melted until the mould is filled. The electrode is usually of a highly refractory metal such as tungsten or of graphite or metal carbide, and in order to minimize vaporization of the electrode, melting is usually carried out under a low pressure of inert gas (argon or helium).

21.6 Electron Beam Melting

The electron beam melting technique has been developed for reactive and refractory metals. Electron beam melting is a process which utilizes the energy of highly accelerated electrons to produce thermal energy generated in the material itself for melting and refining. This is a relatively new means for the direct melting of high melting point metals into ingots of high purity. The heat source is the beam of high velocity electrons. Vacuum is needed for the production and operation of the electron beam. High power electron gun is required for generating the electron

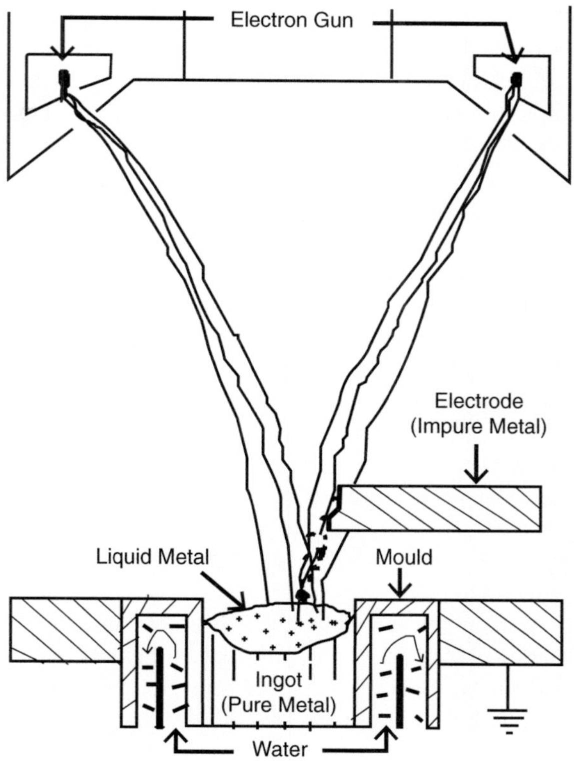

Fig. 21.5 Electron beam melting unit

beam. When a highly accelerated beam of electrons impinges on material, the kinetic energy of the electrons is transferred to the lattice of the material on impact, and its temperature rises gradually due to the thermal energy generated in the material and that melts. The electrons from an electron gun are accelerated and focused onto the metal to be melted by a high voltage potential between the gun and the metal in a high vacuum chamber. The molten metal is collected and allowed to solidify in a water-cooled copper mould to minimize contamination. The electron guns are basically horizontal, ring-shaped, single-run tungsten filaments.

The electron beam melting unit is shown in Fig. 21.5. Impure metal acts as electrode, due to the electron beam impingement on that, metal melts and droplets of molten metal fall and are collected at the water-cooled mould. Many impurities are vaporized from the metal by the combination of high temperature and low pressure. Material can be in the molten state for long periods to eliminate the volatile impurities, as the vacuum is maintain at 10^{-3} mm of Hg. These factors help in obtaining a purer ingot than in the case of vacuum arc melting. Tantalum, niobium, zirconium and hafnium are processed by electron beam melting technique. In addition, the purity of metals such as molybdenum, tungsten and titanium can be improved by this technique.

This form of melting not only consolidates metals of high melting points into ingots, but also causes a great decrease in the impurity content. The rate of purification depends on the initial impurity content of the charge material. Many impurities are separated from the metal by the combination of high temperature and low pressure (under vacuum). A vacuum of higher than 10^{-3} mm of Hg is used, which leads to selective distillation of volatile impurities. In this way the removal of higher volatility impurities by direct evaporation is made possible. This important fact of the purification process combined with the water-cooled non-contaminating casting mould brings about the basic improvement and the higher purities ascribed to electron beam melting. Due to vacuum, contamination by gases during melting is also avoided. The degree of impurity removal is controlled by the vapour pressure of the impurity elements and relative to the vapour pressure of the metal which is treated at the temperature of molten metal.

The electron beam melting and purification process owes its success to the ability to fuse any material known to undergo a solid/liquid transition and maintain the same in the fused state for as long as is necessary to effect the purification required. In this manner, related purification takes place, either by direct evaporation of impurities or the products of reaction occurring in the fused metal at elevated temperatures.

Maintaining the molten state of metal, for a longer time, is possible in electron beam melting; this is an advantage of this process. There is the realization that the electron beam itself has some truly outstanding advantages as a heat source for melting. Its cleanness, the lack of a real upper temperature limit, the realization that vacuum, its natural environment, and a most desirable atmosphere all are important contributing factors. Due to this, more effective refining can be done as compared to the vacuum arc re-melting. However, there are some limitations of the process to the extent of purification possible. In order to characterize the purification achievable in the electron beam melting and refining process, the following items must be considered:

(1) Vaporization of sub-oxides: the ability to remove a specific oxide by the process depends on the relative vapour pressures of an oxide to its respective metal. If the oxide has higher vapour pressure than the metal, then that can be removed easily.
(2) Evolution and removal of gases: removal of gases in general, and the interstitial O, H and N in particular. Removals of interstitial gases are essentially a function of time, which affects the melt rate and economic factors.
(3) Carbon-oxygen reaction: carbon to combine with oxygen, the product (CO gas) can be removed.
(4) Vaporization of metallic impurities: direct removal of higher volatility metallic elements from the metal to be melted and refined.
(5) Floating of non-melting insoluble compounds: the floating to the surface of the melt of all insolubles in the melt materials and inert inclusions which are neither decomposed nor in any way affected in the processing.

(6) Composition segregation, i.e. zone solidification effects: besides melting and resolidifying of the molten metal, the condition of directional solidification exists, both composition segregation and a quasi zone refining effect take place.
(7) Potential sources of contamination: nature of furnace walls and deposits on them, the electron emitter cathode material are the sources of contamination.

The electron beam melting processes are used for metals like tantalum, niobium, molybdenum, tungsten, zirconium, hafnium and other non-ferrous metals and alloys (e.g. Co, V, U, Cu and Ni). After the electron beam melting processes, hydrogen, nitrogen and oxygen content in metal comes down to 1–2 ppm, 2–10 ppm and 3–20 ppm respectively.

Appendix

Physical properties of metals

Metal	Symbol	Atomic weight	Specific gravity	Melting point (°C)	Boiling point (°C)
Aluminium	Al	26.981	2.7	660	2520
Antimony	Sb	121.7	6.68	630.5	1590
Arsenic	As	74.921	5.7	Volatilizes	610 (Sublimes)
Barium	Ba	137.3	3.5	710	1770
Beryllium	Be	9.012	1.8	1280	2450
Bismuth	Bi	208.98	9.8	271.3	1530
Boron	B	10.81	2.3	2030	2550 (Sublimes)
Cadmium	Cd	112.4	8.65	320.9	767
Calcium	Ca	40.08	1.55	843	1484
Chromium	Cr	51.996	7.19	1857	2672
Cobalt	Co	58.933	8.9	1493	2930
Copper	Cu	63.54	8.96	1083	2580
Gold	Au	196.97	19.32	1063	2860
Hafnium	Hf	178.4	13.29	2227	4600
Iron	Fe	55.84	7.9	1539	2860
Lead	Pb	207.2	11.37	327.4	1750
Lithium	Li	6.94	0.53	181	1324
Magnesium	Mg	24.305	1.74	651	1107
Manganese	Mn	54.938	7.3	1244	2150
Molybdenum	Mo	95.9	10.22	2610	5560
Neptunium	Np	237.05	–	637	–
Nickel	Ni	58.69	8.88	1445	3075
Niobium	Nb	92.906	8.57	2415	3300
Palladium	Pd	106.42	12.0	1552	2940
Platinum	Pt	195	21.45	1769	4100
Plutonium	Pu	239.06	19.5	639	3235
Potassium	K	39.09	0.86	63.2	779

(continued)

S.K. Dutta and D.R. Lodhari, *Extraction of Nuclear and Non-ferrous Metals*,
Topics in Mining, Metallurgy and Materials Engineering,
https://doi.org/10.1007/978-981-10-5172-2

(continued)

Metal	Symbol	Atomic weight	Specific gravity	Melting point (°C)	Boiling point (°C)
Rhodium	Rh	102.91	12.4	1966	3700
Scandium	Sc	44.956	3.0	1538	2870
Silicon	Si	28.085	2.37	1414	2287
Silver	Ag	107.87	10.5	960.8	2200
Sodium	Na	22.989	0.97	97.8	883
Tantalum	Ta	180.95	16.6	2996	6000
Thorium	Th	232.04	11.5	1750	4200
Tin	Sn	118.71	7.29	232	2270
Titanium	Ti	47.88	4.54	1668	3285
Tungsten	W	183.8	19.32	3410	5927
Uranium	U	238.03	19.05	1132	3813
Vanadium	V	50.941	6.11	1900	3000
Zinc	Zn	65.39	7.14	419.5	907
Zirconium	Zr	91.224	6.45	1852	4400

Some Thermodynamic Data*

It is important that a metallurgist has access to thermodynamic data of the systems of interest at temperatures of relevance. Some of the classical compilations are listed below:

1. *Metallurgical Thermo-Chemistry*, O. Kubaschewski and C.B. Alcock, Pergamon Press, Oxford (1989).
2. *Materials Thermochemistry*, O. Kubaschewski, C.B. Alcock and P.J. Spencer, Pergamon Press (1993).
3. *Thermochemical Properties of Inorganic Substances*, I. Barin and O. Knacke, Springer-Verlag, Berlin (1973, Supplement 1977).
4. *JANAF Thermochemical Tables*, N.B.S., Michigan (1965 – 68, Supplement 1974-75). 4. Selected Values of Thermochemical Properties of Metals and Alloys, R. Hultgren, R.L. Orr, P.D. Anderson and K.K. Kelley, John Wiley & Sons, NY (1963).
5. *Metallurgical Thermodynamics, Kinetics & Numericals*, S.K. Dutta and A.B. Lele, S. Chand & Co Ltd, New Delhi (2012).
6. *Smithells Metals Reference Book*, Eds by E.A. Brandes & G.B. Brook, 7th Ed, Reed Educational and Professional Publishing Ltd, Oxfort (1992).

S.K. Dutta and D.R. Lodhari, *Extraction of Nuclear and Non-ferrous Metals*,
Topics in Mining, Metallurgy and Materials Engineering,
https://doi.org/10.1007/978-981-10-5172-2

Bibliography

1. *An Introduction to the Study of Fuel*, J.C. Macrae, Elsevier Publishing Company, London (1966)
2. *Electro-metallurgy of Steel and Ferro-alloys*, F.P. Edneral, vol-2, Mir Publishers, Moscow (1979)
3. *Extraction of Non-Ferrous Metals*, H.S. Ray, R. Sridhar and K.P. Abraham, Affiliated East-West Press Pvt Ltd, New Delhi (1987)
4. *Metallurgy of the Non-Ferrous Metals*, W.H. Dennis, Sir Isaac Pitman & Sons Ltd, London (1966)
5. *Non-Ferrous Metallurgy, N. Sevryukov*, Mir Publishers, Moscow, (1975)
6. *Nuclear Power, D. Voskoboinik*, Foreign Languages Publishing House, Moscow, (1959)
7. *Nuclear Reactor Engineering*, S. Glasstone and A. Sesouske, CBS Publisher & Distrubutors, New Delhi (1986)
8. *Nuclear Reactor Fuel Elements (Metallurgy and Fabrication)*, Ed. A.R. Kaufmann, Interscience Publisher, London (1958)
9. Proc. of Seminar on *Nuclear Metallurgy*, org by IIM Bombay Chapter and IE(I), Maharashtra State Centre, India (1978)
10. *Production of Ferro-alloys*, A. Riss and Y. Khodorovsky, Foreign Languages Publishing House, Moscow (1959)
11. *Rare Metals Handbook*, Ed by C.A. Hampel, Reinhold Publishing Co, New York (1967)
12. *Vacuum Metallurgy*, Eds by O. Winkler and R. Bakish, Elsevier Publishing Co, Amsterdam, Netherlands (1971)

S.K. Dutta and D.R. Lodhari, *Extraction of Nuclear and Non-ferrous Metals*,
Topics in Mining, Metallurgy and Materials Engineering,
https://doi.org/10.1007/978-981-10-5172-2

MIX
Papier aus verantwortungsvollen Quellen
Paper from responsible sources
FSC® C105338

If you have any concerns about our products,
you can contact us on
ProductSafety@springernature.com

In case Publisher is established outside the EU,
the EU authorized representative is:
Springer Nature Customer Service Center GmbH
Europaplatz 3, 69115 Heidelberg, Germany

Printed by Libri Plureos GmbH
in Hamburg, Germany